Information Circular 9527

National Survey of the Mining Population
Part I: Employees

Linda Jansen McWilliams, Patricia J. Lenart, Jamie L. Lancaster, John R. Zeiner, Jr.

DEPARTMENT OF HEALTH AND HUMAN SERVICES
Centers for Disease Control and Prevention
National Institute for Occupational Safety and Health
Office of Mine Safety and Health Research
Pittsburgh, PA • Spokane, WA

June 2012

This document is in the public domain and may be freely copied or reprinted.

Disclaimer

Mention of any company or product does not constitute endorsement by the National Institute for Occupational Safety and Health (NIOSH). In addition, citations to Web sites external to NIOSH do not constitute NIOSH endorsement of the sponsoring organizations or their programs or products. Furthermore, NIOSH is not responsible for the content of these Web sites. All Web addresses referenced in this document were accessible as of the publication date.

Ordering Information

To receive documents or other information about occupational safety and health topics, contact NIOSH at

>Telephone: **1–800–CDC–INFO** (1–800–232–4636)
>TTY: 1–888–232–6348
>e-mail: cdcinfo@cdc.gov

>or visit the NIOSH Web site at **www.cdc.gov/niosh**.

For a monthly update on news at NIOSH, subscribe to NIOSH *eNews* by visiting **www.cdc.gov/niosh/eNews**.

DHHS (NIOSH) Publication No. 2012–152

June 2012

SAFER • HEALTHIER • PEOPLE™

Section Contents

- Employee Statistics for All Mines

- Employee Statistics for Coal Mines

- Employee Statistics for Metal Mines

- Employee Statistics for Nonmetal Mines

- Employee Statistics for Stone Mines

- Employee Statistics for Sand and Gravel Mines

Contents

Abstract 1
Introduction 2
Background and Overview 2
Survey Materials 3
    The Paper Questionnaire 3
    The Internet Questionnaire 4
Sample Design and Selection 4
    Definition of the Target Population 4
    Construction of the Sampling Frame 5
    Stratification Guidelines... 5
Sampling Plans 15
Data Collection 19
    Survey Packet 19
    Survey Promotion 20
    Follow-up Contacts 21
Data Imputation and Statistical Weighting Procedures.. 21
    Data Imputation 21
    Data Weighting, Estimation, and Variance Estimation 22
Lessons Learned... 22
Survey Results 23
Employee Job Titles 27
Statistical Analysis... 28
Employee Statistics for All Mines 29
Employee Statistics for Coal Mines 55
Employee Statistics for Metal Mines.. 73
Employee Statistics for Nonmetal Mines 91
Employee Statistics for Stone Mines 111
Employee Statistics for Sand and Gravel Mines 131
Acknowledgements 149
References 151
Appendices 153
Appendix A. Questionnaire Booklet... 155

v

Appendix B. Questions and Answers Brochure 179
Appendix C. MSHA Form 7000-2: Quarterly Mine Employment and Coal Production Report 183
Appendix D. Standard Industrial Classifications (SIC) for Active Mines in 2007 187
Appendix E. Stratification and Sample Size Guidelines 191
Appendix F. Sample Size Allocation Using MSHA Data from the Second Quarter of 2002 197
Appendix G. Critical Items from the Questionnaire 209
Appendix H. Job Titles as Submitted by Survey Respondents 215
Appendix I. Glossary 243

Figures

Figure 1. Map of Active Coal Mining Operations for 2007. 7
Figure 2. Map of Active Metal Mining Operations for 2007. 8
Figure 3. Map of Active Nonmetal Mining Operations for 2007. 9
Figure 4. Map of Active Stone Mining Operations for 2007. 10
Figure 5. Map of Active Sand and Gravel Mining Operations for 2007. 11
Figure 6. Density Map for Mine Operator Employees for 2007. 12
Figure 7. Density Map for Underground Mine Operator Employees for 2007. 13
Figure 8. Density Map for Surface Mine Operator Employees for 2007. 14
Figure 9. Weighted Response Rates by Sector and Mine Type. 26
Figure 10. Education Level of Employees at All Mines. 35
Figure 11. Race of Employees at All Mines. 36
Figure 12. Primary Work Location of Employees at All Mines. 37
Figure 13. Occupational Categories of Employees at All Mines. 53
Figure 14. Education Level of Employees at Coal Mines. 61
Figure 15. Race of Employees at Coal Mines. 62
Figure 16. Primary Work Location of Employees at Coal Mines. 63
Figure 17. Occupational Categories of Employees at Coal Mines. 72
Figure 18. Education Level of Employees at Metal Mines. 79
Figure 19. Race of Employees at Metal Mines. 80
Figure 20. Primary Work Location of Employees at Metal Mines. 81
Figure 21. Occupational Categories of Employees at Metal Mines. 89
Figure 22. Education Level of Employees at Nonmetal Mines. 97
Figure 23. Race of Employees at Nonmetal Mines. 98
Figure 24. Primary Work Location of Employees at Nonmetal Mines. 99
Figure 25. Occupational Categories of Employees at Nonmetal Mines. 109
Figure 26. Education Level of Employees at Stone Mines. 117
Figure 27. Race of Employees at Stone Mines. 118
Figure 28. Primary Work Location of Employees at Stone Mines. 119
Figure 29. Occupational Categories of Employees at Stone Mines. 129
Figure 30. Education Level of Employees at Sand and Gravel Mines. 137
Figure 31. Race of Employees at Sand and Gravel Mines. 138
Figure 32. Primary Work Location of Employees at Sand and Gravel Mines. 139
Figure 33. Occupational Categories of Employees at Sand and Gravel Mines. 147

Tables

Table 1. Sample Allocation for Underground Coal Mines 16
Table 2. Sample Allocation for Surface Coal Mines 16
Table 3. Sample Allocation for Underground Metal Mines 16
Table 4. Sample Allocation for Surface Metal Mines 17
Table 5. Sample Allocation for Underground Nonmetal Mines 17
Table 6. Sample Allocation for Surface Nonmetal Mines 17
Table 7. Sample Allocation for Underground Stone Mines 18
Table 8. Sample Allocation for Surface Stone Mines 18
Table 9. Sample Allocation for Sand and Gravel Mines 18
Table 10. Number of Mines in the Final Sample by Sector, Type, and Reporting Week 20
Table 11. Summary of Final Results for All Sampled Mines 24
Table 12. Number of Completed Surveys by Mode 24
Table 13. Summary of Ineligible Mines by Sector 25
Table 14. Summary of Refusal by Mine Sector and Type of Refusal 25
Table 15. National Estimates of Mines and Mine Employees in Spring/Summer 2008 27
Table 16. Demographic Characteristics of Employees at All Mines 33
Table 17. Occupational Characteristics of Employees at All Mines 34
Table 18. Estimated Number of Administration/Professional Employees at All Mines . 38
Table 19. Estimated Number of Maintenance Employees at All Mines 43
Table 20. Number of Miscellaneous Employees at All Mines 44
Table 21. Estimated Number of Production Employees at All Mines 45
Table 22. Estimated Number of Service and Utility Employees at All Mines 50
Table 23. Demographic Characteristics of Employees at Coal Mines 59
Table 24. Occupational Characteristics of Employees at Coal Mines 60
Table 25. Estimated Number of Administration/Professional Employees at Coal Mines 64
Table 26. Estimated Number of Maintenance Employees at Coal Mines 67
Table 27. Number of Miscellaneous Employees at Coal Mines 68
Table 28. Estimated Number of Production Employees at Coal Mines 68

Table 29. Estimated Number of Service and Utility Employees at Coal Mines............ 70
Table 30. Demographic Characteristics of Employees at Metal Mines...................... 77
Table 31. Occupational Characteristics of Employees at Metal Mines 78
Table 32. Estimated Number of Administration/Professional Employees at Metal Mines 82
Table 33. Estimated Number of Maintenance Employees at Metal Mines 85
Table 34. Number of Miscellaneous Employees at Metal Mines 86
Table 35. Estimated Number of Production Employees at Metal Mines...................... 86
Table 36. Estimated Number of Service and Utility Employees at Metal Mines 88
Table 37. Demographic Characteristics of Employees at Nonmetal Mines................. 95
Table 38. Occupational Characteristics of Employees at Nonmetal Mines 96
Table 39. Estimated Number of Administration/Professional Employees at Nonmetal Mines 100
Table 40. Estimated Number of Maintenance Employees at Nonmetal Mines........... 103
Table 41. Number of Miscellaneous Employees at Nonmetal Mines 104
Table 42. Estimated Number of Production Employees at Nonmetal Mines 104
Table 43. Estimated Number of Service and Utility Employees at Nonmetal Mines... 107
Table 44. Demographic Characteristics of Employees at Stone Mines...................... 115
Table 45. Occupational Characteristics of Employees at Stone Mines 116
Table 46. Estimated Number of Administration/Professional Employees at Stone Mines 120
Table 47. Estimated Number of Maintenance Employees at Stone Mines................. 123
Table 48. Number of Miscellaneous Employees at Stone Mines 124
Table 49. Estimated Number of Production Employees at Stone Mines 124
Table 50. Estimated Number of Service and Utility Employees at Stone Mines......... 127
Table 51. Demographic Characteristics of Employees at Sand and Gravel Mines 135
Table 52. Occupational Characteristics of Employees at Sand and Gravel Mines..... 136
Table 53. Estimated Number of Administration/Professional Employees at Sand and Gravel Mines.. 140
Table 54. Estimated Number of Maintenance Employees at Sand and Gravel Mines 142
Table 55. Number of Miscellaneous Employees at Sand and Gravel Mines.............. 143
Table 56. Estimated Number of Production Employees at Sand and Gravel Mines... 143

Table 57. Estimated Number of Service and Utility Employees at Sand and Gravel Mines 145

Table E-1. Half-Length of 95% Confidence Intervals in Percentage Points for Various Percentages Being Estimated for Domains of Various Sizes with Various Design Effects.. 196

Table F-1. Sample Allocation for Underground Coal Mines 199

Table F-2. Sample Allocation for Surface Coal Mines 200

Table F-3. Sample Allocation for Underground Metal Mines 201

Table F-4. Sample Allocation for Surface Metal Mines... 202

Table F-5. Sample Allocation for Underground Nonmetal Mines 203

Table F-6. Sample Allocation for Surface Nonmetal Mines 204

Table F-7. Sample Allocation for Underground Stone Mines 205

Table F-8. Sample Allocation for Surface Stone Mines 206

Table F-9. Sample Allocation for Sand and Gravel Mines.. 207

Acronyms and Abbreviations

CI	Confidence Interval
DEFF	Design Effect
DSU	Data suppressed
FPC	Finite population corrected
FTE	Full-time Equivalent
IC	Information Circular
LCL	Lower Confidence Limit
MIPS	Mining Industry Population Survey
MSHA	Mine Safety and Health Administration
NA	Not applicable
NIOSH	National Institute for Occupational Safety and Health
OMB	Office of Management and Budget
OMSHR	Office of Mine Safety and Health Research
SIC	Standard Industrial Classification
UCL	Upper Confidence Limit

Definition of Terms

Confidence Interval:	An interval that gives an estimated range of values which is likely to include an unknown population parameter, the estimated range being calculated from a given set of sample data
Jackknife Repeated Replication:	A commonly used resampling approach to variance estimation
Lower Confidence Limit:	The lower bound of a confidence interval
National Estimate:	A weighted statistical calculation which uses the results from a probability sample survey to estimate a national number
Survey Count:	The actual number of responses obtained from the National Survey of the Mining Population
Upper Confidence Limit:	The upper bound of a confidence interval

National Survey of the Mining Population
Part I: Employees

Linda Jansen McWilliams, Patricia J. Lenart, Jamie L. Lancaster, John R. Zeiner, Jr.

Office of Mine Safety and Health Research
National Institute for Occupational Safety and Health

Abstract

The National Institute for Occupational Safety and Health (NIOSH) conducted the first comprehensive survey of the U.S. mining population in more than 20 years. The National Survey of the Mining Population captured the current profile of the U.S. mining workforce. Data collection began in March 2008 and continued through August 2008. Randomly selected mining operations in all of the major mining sectors (i.e., coal, metal, nonmetal, stone, and sand and gravel) received the survey and had the option of completing a paper or web-based questionnaire. A total of 737 mining operations returned completed questionnaires and reported data for 9,008 employees.

Two sets of data were collected in this national survey. There were questions about the mining operation, including employee training, work schedules, the use of independent contractor employees, and mine communication and safety systems. The employee questions included demographic and occupational questions about individual employees. The survey sample data were weighted in order to provide national estimates of mine and employee characteristics.

This Information Circular (IC) is published in two parts—"Part I: Employees" presents the employee-level data and "Part II: Mines" presents the mine-level data. Both parts of this IC include an overview of the survey background, development of the survey materials, sample design and sample selection, data collection and processing, statistical weighting, and lessons learned. The survey data are summarized for the overall U.S. mining industry and the five major mining sectors. The information gathered from the survey respondents is being published only as summarized data so that no single mining operation or employee can be identified.

Introduction

Surveillance of occupational injuries, illnesses, and exposures has been an integral part of the work of the National Institute for Occupational Safety and Health (NIOSH) since its creation by the Occupational Safety and Health Act in 1970. Surveillance activities at the Office of Mine Safety and Health Research (OMSHR) are focused on the nation's mining workforce. These surveillance activities make extensive use of data from a number of different national databases. The most frequently used databases are those maintained by the Mine Safety and Health Administration (MSHA). Included are databases of reported employment, accidents/injuries/illnesses, hazardous exposures, coal production, mine inspections, violations and citations, etc. Two of the most commonly used databases are the mine operator and contractor address/employment file and the file listing reports of accidents, injuries, and illnesses.

Analysis of data from the existing MSHA employment and accident/injury/illness databases has been able to meet some, but not all, of the OMSHR surveillance needs. For example, to identify subpopulations in each major mining sector or type of mining operation at risk of adverse health and safety outcomes, OMSHR needs the capability to calculate age-, gender-, and occupation-specific rates of injuries, fatalities, and disease. Additionally, due to the reduced reporting requirements for independent contractors, OMSHR cannot determine the number of contractor employees working separately in metal, nonmetal, stone, or sand and gravel operations. The National Survey of the Mining Population was designed to collect mine- and employee-level information to address these and other data gaps.

Background and Overview

The last national survey targeting the mining workforce, the Mining Industry Population Survey (MIPS), was conducted in 1986 by the U.S. Bureau of Mines (USBM) in the U.S. Department of the Interior. The mining industry has experienced many changes since the MIPS was conducted, and its data are too outdated to be considered useful for surveillance on the current mining workforce. In addition, the MIPS did not include any information on independent contractor employees. Therefore, the National Institute for Occupational Safety and Health, Office of Mine Safety and Health Research conducted this survey to provide updated demographic and occupational information on the mining workforce. The National Survey of the Mining Population collected information from each of the five major mining sectors (coal, metal, nonmetal, stone, and sand and gravel). The survey's main objectives were to:

- Collect basic information about mining operations.
- Establish the demographic and occupational characteristics of mine operator employees.
- Estimate the number of independent contractor employees used by mining operations.

Data collection began in March 2008 and continued through August 2008. A survey packet was mailed to each sampled mining operation. Respondents were given the option of completing a paper questionnaire or using a web questionnaire. Two sets of data were collected in this survey. The mine questions included items about the mining operations, communication and safety systems, and the mine's use of independent contractor employees. The employee questions included demographic and occupational questions about individual employees. The survey's employee-level data will be used by OMSHR to determine the accident rates for various demographic and occupational categories as well as provide information that will be used to improve the safety and health of miners.

This Information Circular (IC) is published in two parts—"Part I: Employees" presents the employee-level data and "Part II: Mines" presents the mine-level data. The employee and mine data are summarized for the overall U.S. mining industry and the major mining sectors. In addition, the data in the Mines IC is stratified by underground and surface for the coal, metal, nonmetal, and stone sectors. The information gathered from the survey respondents is being published only as summarized data so that no single mining operation or employee can be identified. The intent of this IC is to present the methodology used to design and conduct the survey and to provide up-to-date information about U.S. mining operations and their employees.

Survey Materials

A survey packet was developed which contained a cover letter, a questionnaire booklet with employee sampling instructions (Appendix A), directions for accessing the Internet version of the questionnaire, a Questions and Answers (Q&A) brochure (Appendix B), and a stamped, self-addressed return envelope.

The Paper Questionnaire

Each survey paper questionnaire booklet was personalized with a box at the top of page 1 which included: the mine ID number, the mine name, the reporting week (date), and a "submit-by" date. The Questionnaire Overview section presented general instructions and guidelines for completing the survey. The survey consisted of five parts as summarized below:

- Mine Questions—This first part of the questionnaire included sections on: Training; Other Languages; Work Schedules for Production Workers, Production Support Workers, and Preparation Plant/Mill Workers; Shift Work for these same three types of workers; Independent Contractor Employees; and Safety, Communication, and Rescue Measures.
- Employee Selection Instructions—This page contained step-by-step instructions for selecting the sample of employees to be included in the Employee Questions. Personalized mine information was preprinted at the top of this page, including: the mine ID number, the mine name, the reporting week (date), the range of the estimated number of employees working at the mine, a "start-with" number and a "take-every" number for selecting employees from the mine's employee roster.
- Instructions for Employee Questions—This two-page section of the questionnaire provided item-by-item explanations for the Employee Questions.

- Employee Questions—These items were formatted as a fold-out answer form. The sections included: Regular Job Title, Mining Experience, Number of Hours Worked During the Reporting Week, Primary Work Location, Gender, Race, Ethnicity, Birth Year, and Education Level. Two pages of the form were included, with the first page containing lines for reporting up to 15 employees and the second page containing lines for reporting up to 14 additional employees, or a maximum of 29 sampled employees.
- Final Questions and Comments—This two-page section of the questionnaire included: questions for reporting unusual events or circumstances at the mine during the designated reporting week; the date the questionnaire was completed; the name, title and telephone number of the company representative who should be contacted regarding questionnaire completion; space for entering comments or explanations related to specific responses; and mailing instructions.

The Internet Questionnaire

Beginning in October 2004, a pilot study was conducted to evaluate the recruitment materials, questionnaire, and survey procedures developed for the nationwide survey of the mining population. This study allowed OMSHR to explore the feasibility of developing a web-based version of the questionnaire. The pilot study debriefing interview contained several questions to determine whether the mine had access to the Internet and how convenient this would be for completion of the questionnaire. The majority of respondents indicated that an Internet connection was available at their mine and more than 50 percent reported preferring an electronic response option. Thus, for the National Survey of the Mining Population, a web-based survey was made available. The survey contractor developed the web survey, including programming of the administrative interface, Section 508 compliance, data validation, quality assurance, and programming of the critical questions.

Sample Design and Selection

Definition of the Target Population

The target population for a survey is the entire set of population units about which the survey data are to be used to make inferences. Establishment surveys such as the National Survey of the Mining Population must delineate the level of the business organization that constitutes the units of the target population. Because hazards vary across mines, the target population for this survey was defined in terms of the individual mining operation.

The target population of mines consisted of active mines in current production. The survey was further restricted to operations that were covered under Title 30 of the U.S. Code, specifically mines whose mineral output was sold or used in commerce. The target population of employees was restricted to those mine employees for whom the mine operator must report hours worked using the MSHA Form 7000-2: *Quarterly Mine Employment and Coal Production*

Report (Appendix C). This includes all direct employees working at the mine, but not contract employees brought in periodically or regularly to perform work at the mine.

There is an important temporal aspect to these definitions for mines and for mine operator employees. Over time, some mines will go in and out of operation. Similarly, employees join the mining labor force and leave the labor force over time. Accordingly, the National Survey of the Mining Population focused on mines in operation during a particular calendar quarter and the current employees of those mines.

Construction of the Sampling Frame

The sampling frame for a survey is the list or mechanism used to enumerate target population members for sample selection purposes. Individual sampling frames for each of the five major mining sectors (see Figures 1–5) were constructed using the 2007 second quarter data released by the Mine Safety and Health Administration, so that the sampling frames would be in sync with the actual time period when data collection would begin (the second quarter of 2008). To ensure that any startup or intermittent mining operations would not be missed, all mines reporting zero employment hours were included in these frames. Any mines with a status of abandoned or abandoned/sealed were excluded from the sampling frames. The Standard Industrial Classification (SIC) for the active coal, metal, nonmetal, stone, and sand and gravel mines used in the sampling frames is presented in Appendix D.

Stratification Guidelines

For the National Survey of the Mining Population, mine-level and employee-level analyses were planned, which required adequate sample sizes of mines and of employees. Because multiple employees were to be sampled from each responding mine, sample size requirements for mine-level analyses tended to drive the total number of mines that needed to be sampled. The sample size for employees was determined by the number of sampled mines responding and the average number of employees sampled per mine.

The competing needs of mine-level analysis versus employee-level analysis required the use of a compromise design that supported the objectives of both types of analyses. For mine-level analyses, the best design was one that selected mines with equal probability, while for employee-level analyses the best design was one that selected mines with probability proportional to the number of employees. The compromise design met both needs by stratifying by the number of employees and then sampling mines with equal probability within strata. Strata associated with large mines (in terms of employment) were given greater selection probabilities than small mines, which would facilitate employee-level analyses by making the employee selection probabilities less variable across mines.

Mine size was an important domain for study at the mine level as well as at the employee level. For example, mines might be more likely to vary in their training procedures based upon employee size. Small mines may be more likely to use trainers from outside the organization, while large mines may be more likely to rely on in-house trainers. Hence, stratifying by the number of employees when sampling mines served an analytic purpose, as well as facilitated the over sampling of large mines needed for employee-level analyses (see Figure 6).

From an analysis standpoint, it was also desirable to control for underground versus surface mines when sampling mines and employees (see Figures 7 and 8). Underground coal mines, in particular, have higher injury and fatality rates than surface mines. There is substantial diversity in the incidence of injuries and fatalities at underground mines versus surface mines across all mining sectors. Nearly one-third of coal and metal mines are underground. Less than ten percent of nonmetal and stone mines are underground and sand and gravel mines are surface only. Stratification by underground mines versus surface mines allows for the control over sample sizes needed for effective comparisons of underground mines to surface mines. A more in-depth discussion of the stratum size formation and sample size guidelines used in this survey can be found in Appendix E.

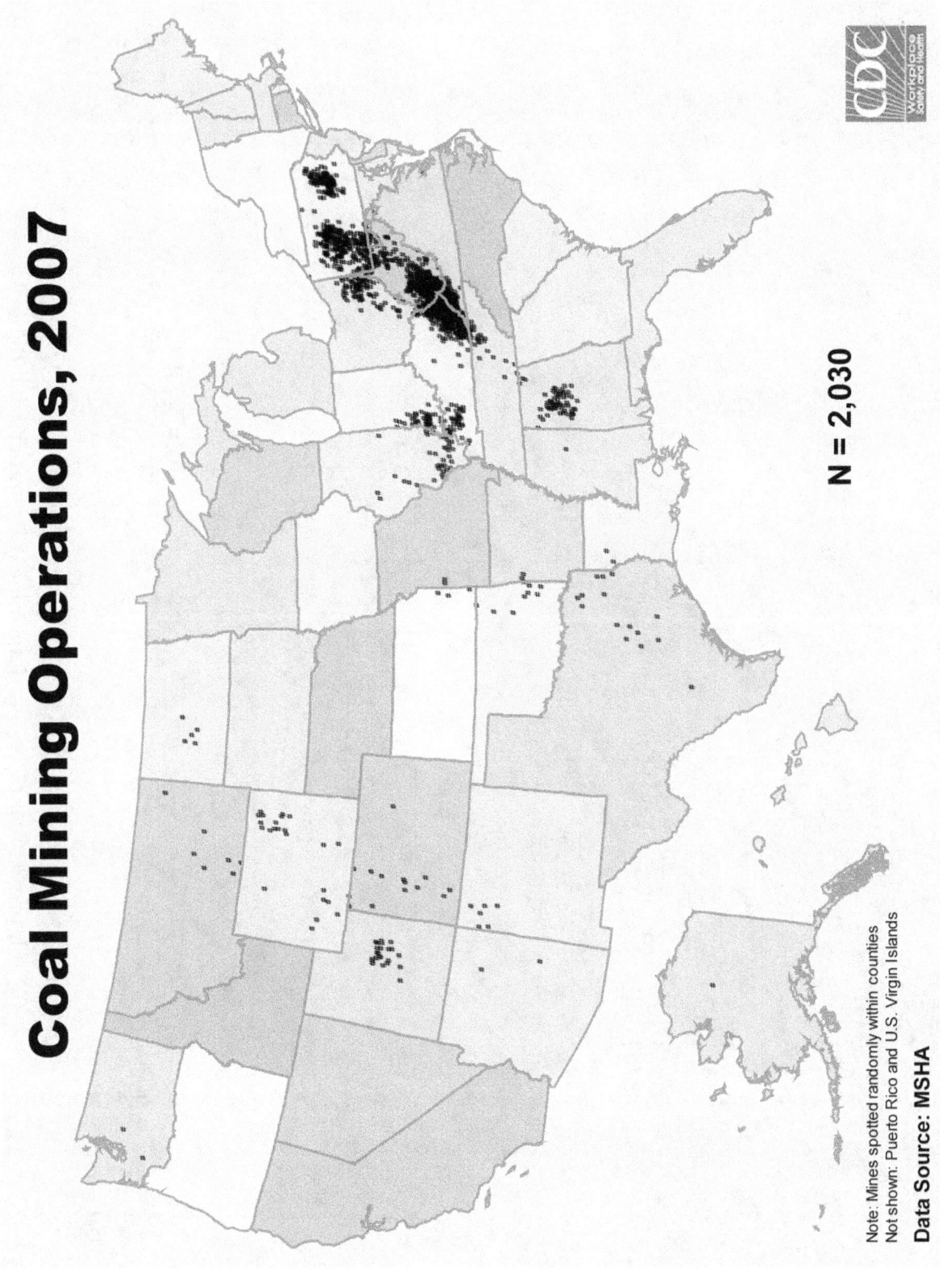

Figure 1. Map of Active Coal Mining Operations for 2007.

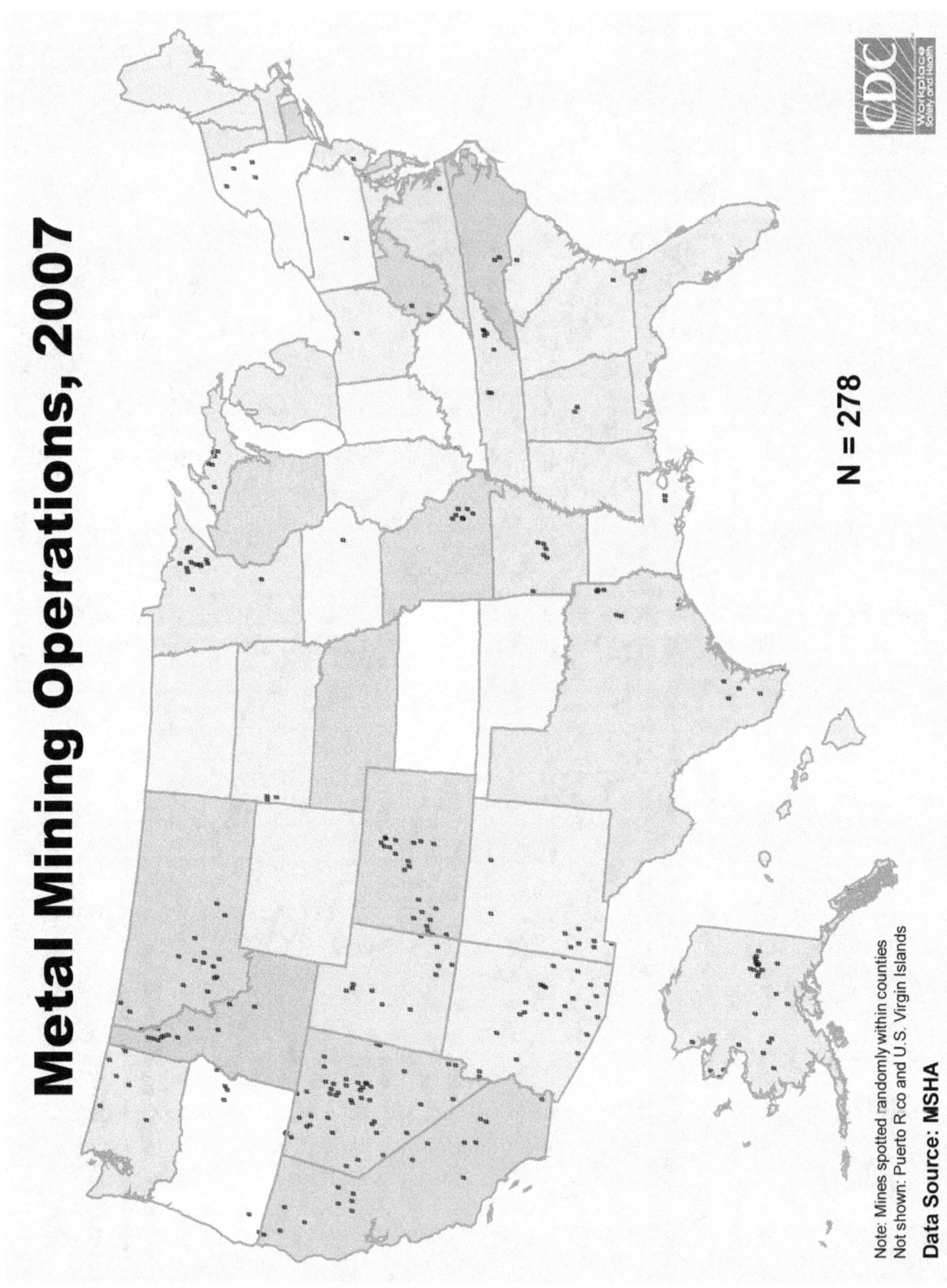

Figure 2. Map of Active Metal Mining Operations for 2007.

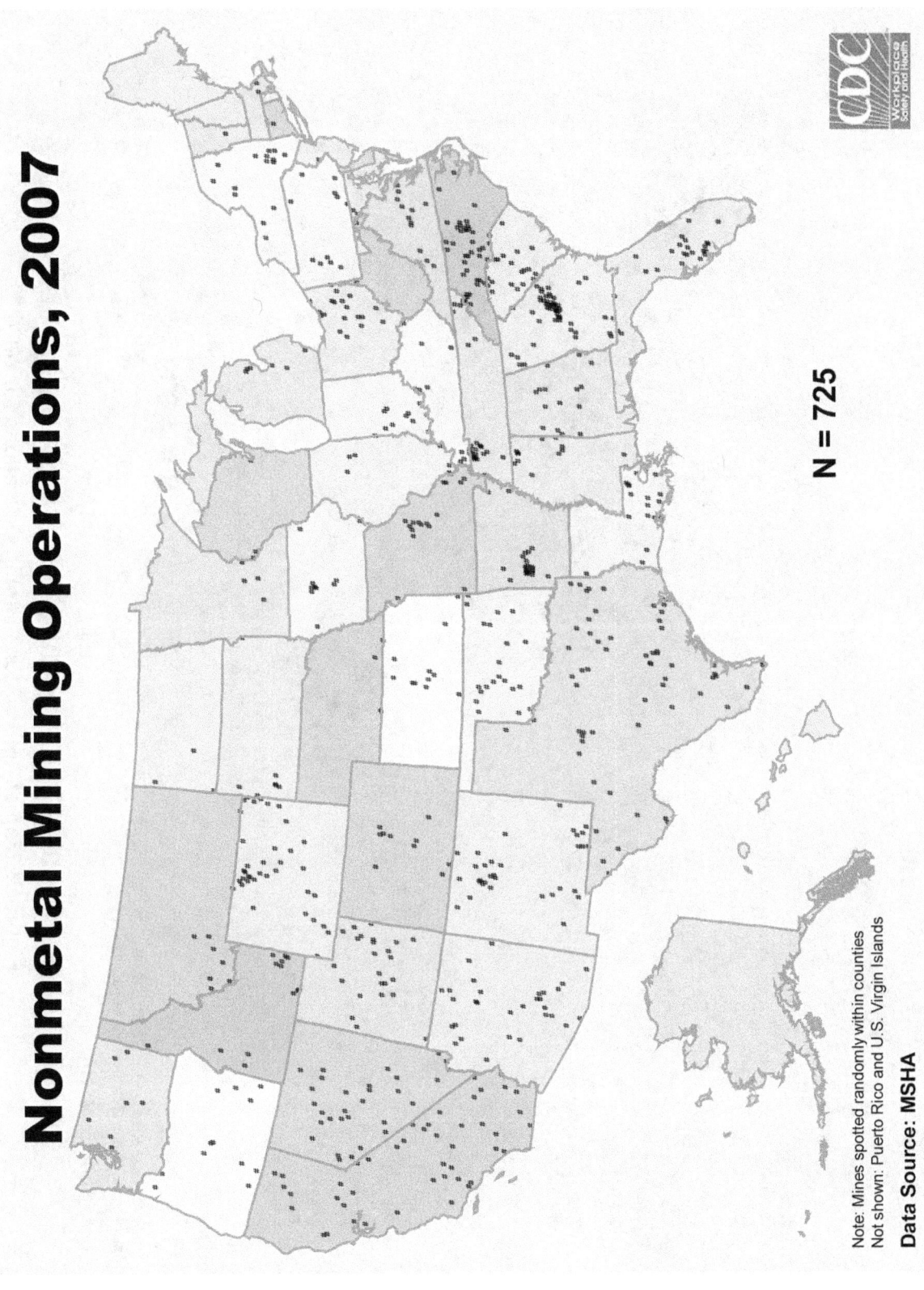

Figure 3. Map of Active Nonmetal Mining Operations for 2007.

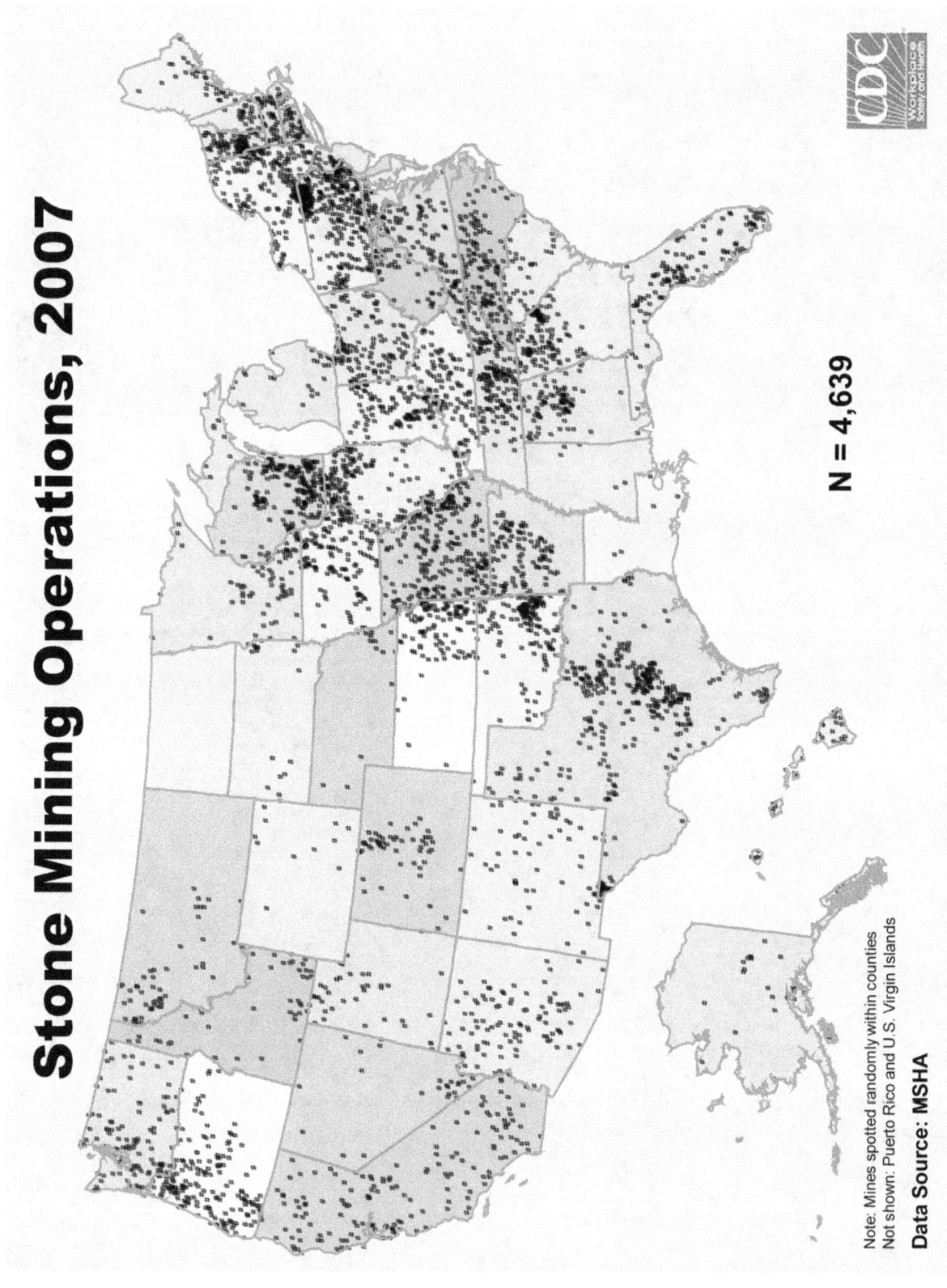

Figure 4. Map of Active Stone Mining Operations for 2007.

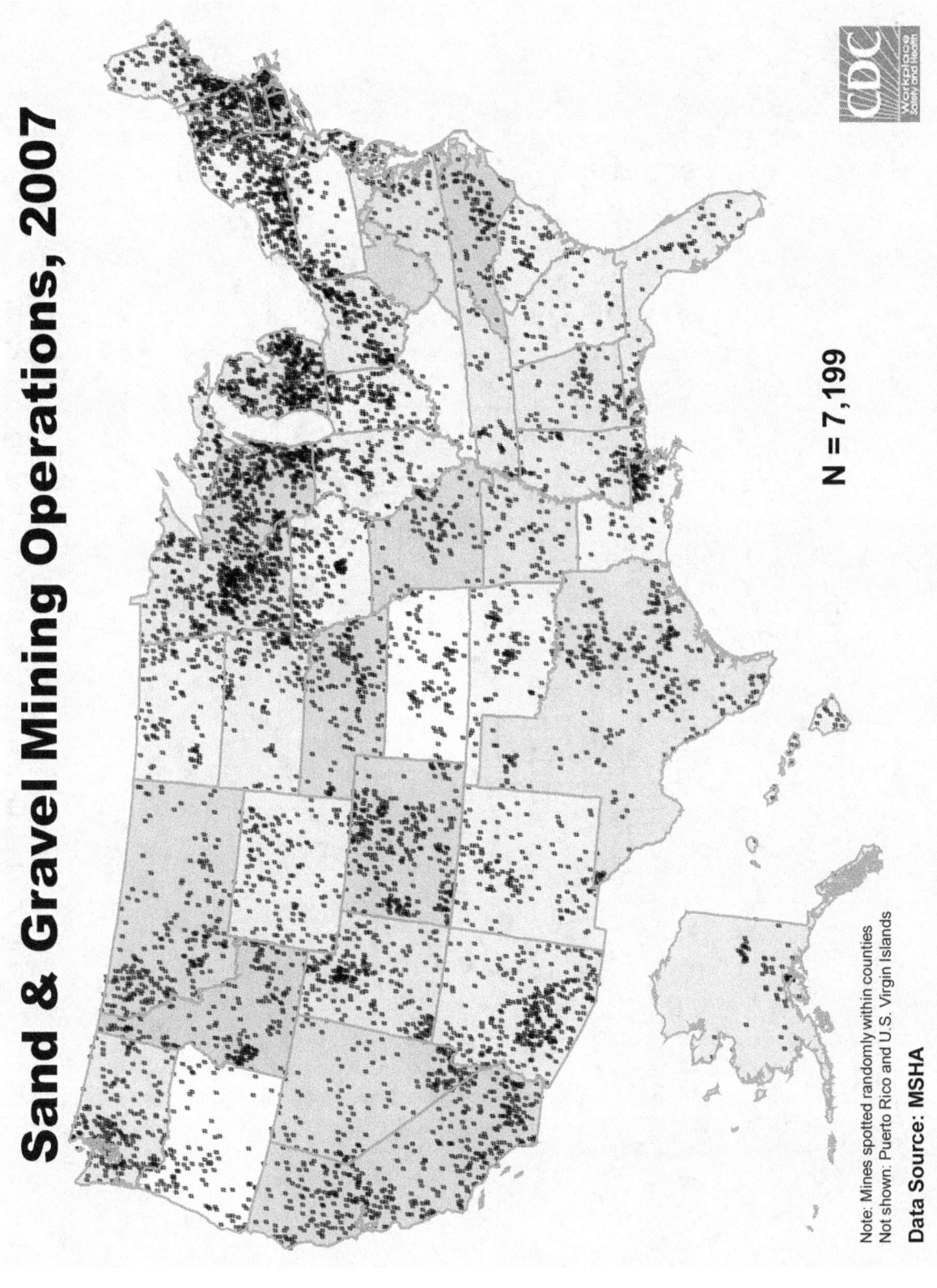

Figure 5. Map of Active Sand and Gravel Mining Operations for 2007.

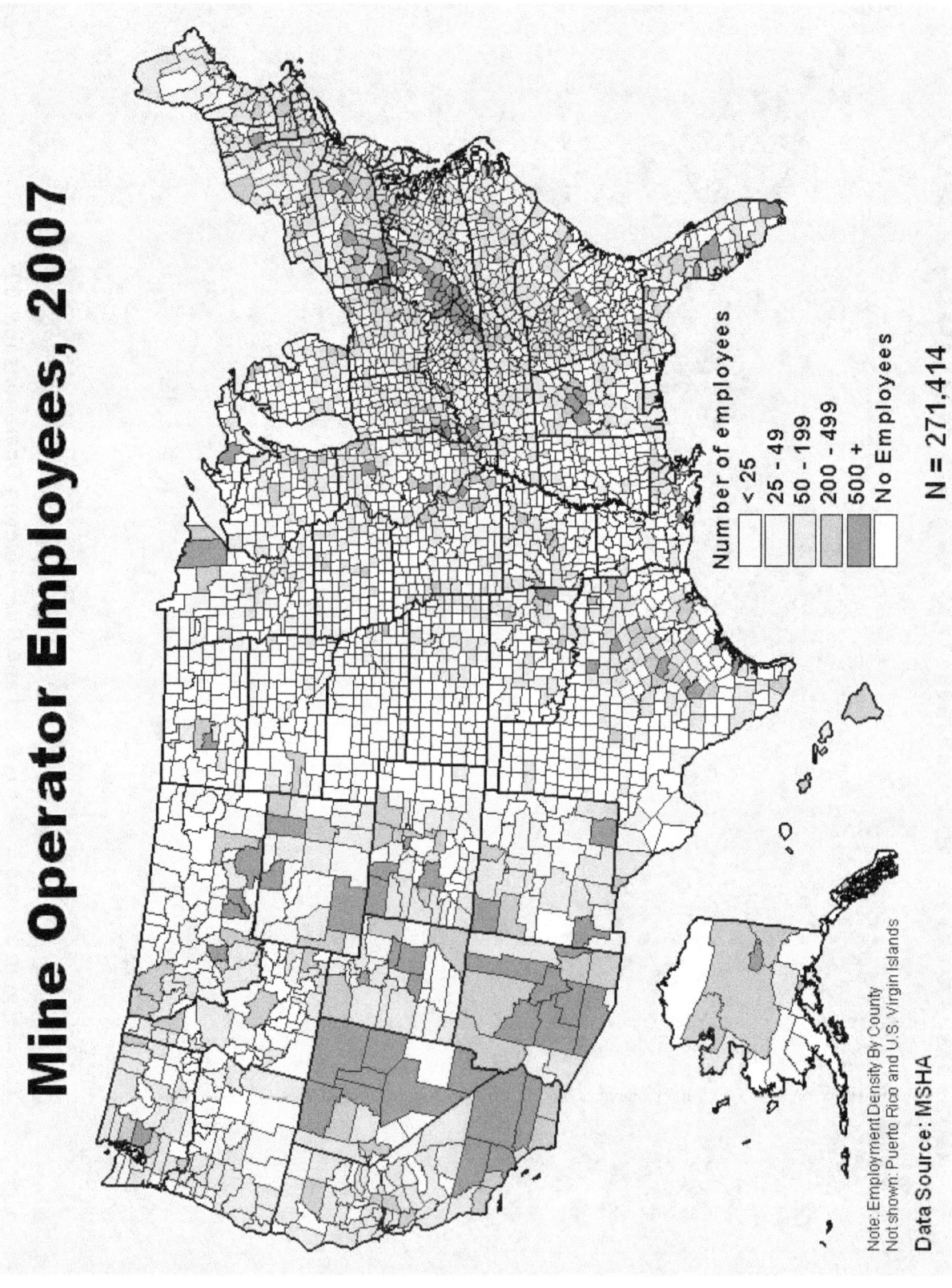

Figure 6. Density Map for Mine Operator Employees for 2007.

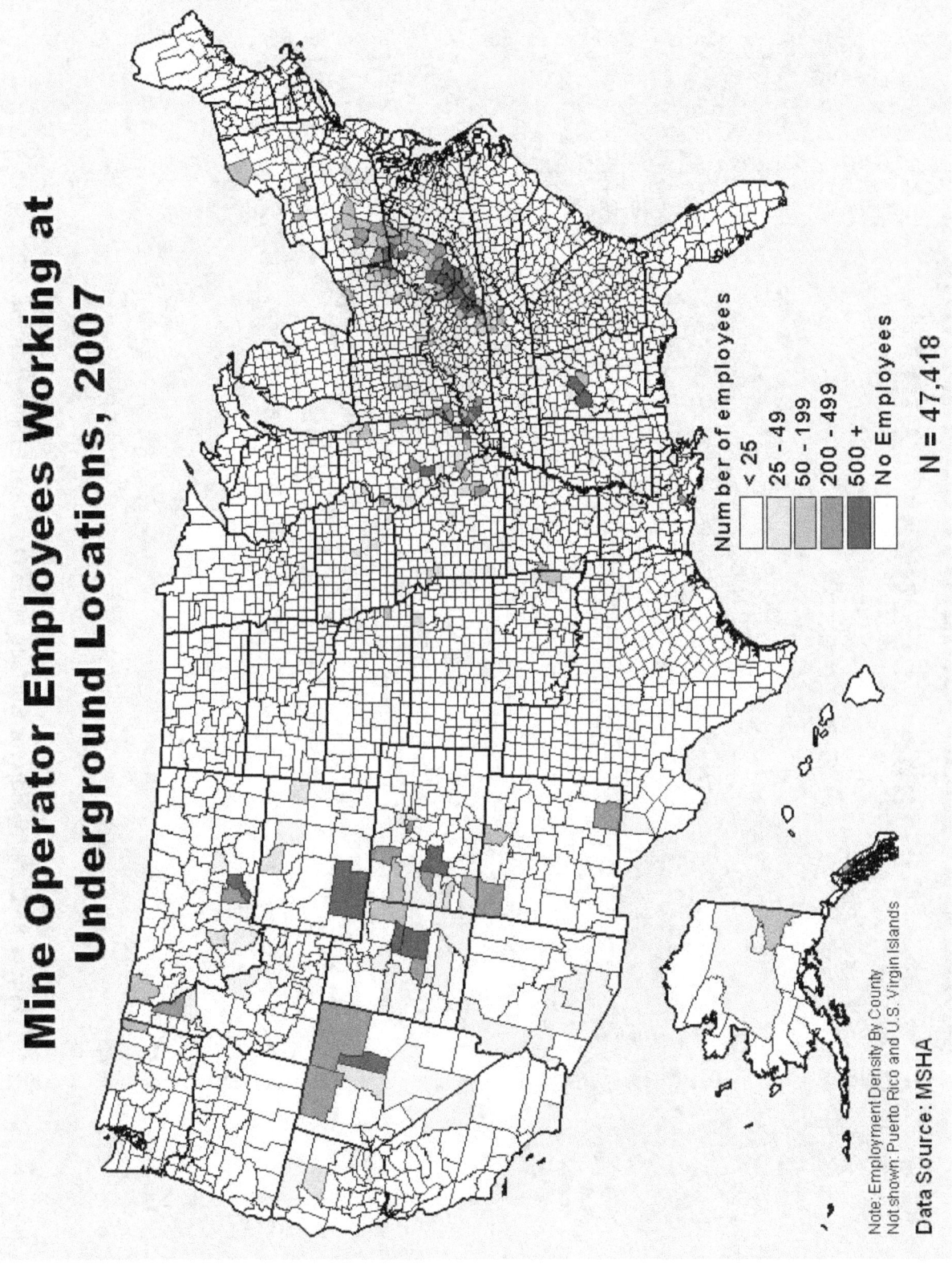

Figure 7. Density Map for Underground Mine Operator Employees for 2007.

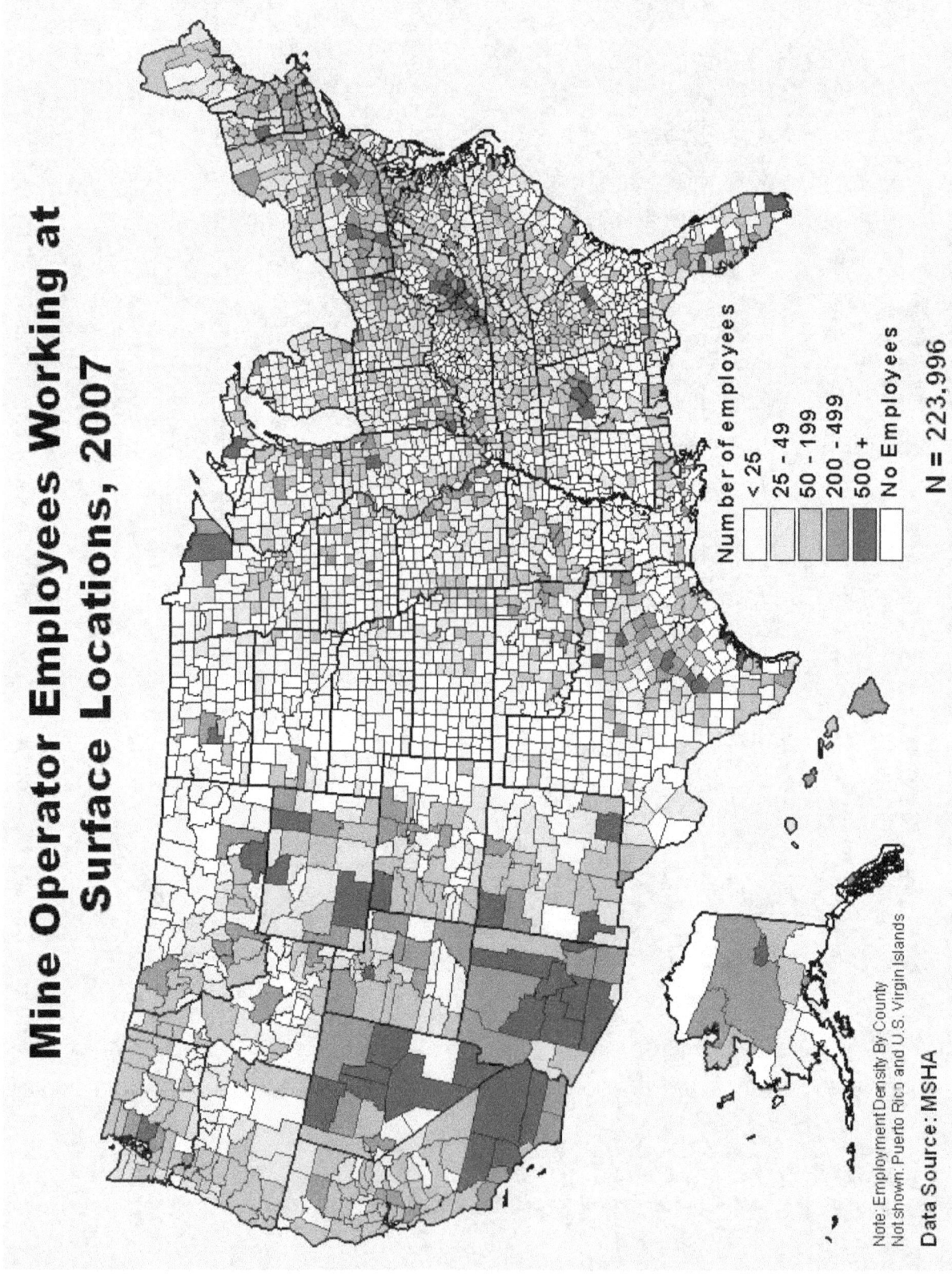

Figure 8. Density Map for Surface Mine Operator Employees for 2007.

Sampling Plans

The original sampling plans were finalized in 2004 after a pretest with eight mining operations. These plans were developed using MSHA data from the second quarter of 2002. The number of actual employees was used to develop these designs rather than the number of full-time equivalent (FTE) employees, because the mine operator would be sampling based upon counts of actual employees, not FTEs. Mines were classified as surface or underground based upon MSHA subunit codes. Mines reporting any employment at underground work locations were classified as underground mining operations.

Because there would actually be two surveys, one for mines and one for employees, the sampling allocation needed to be balanced. An approach that Cochran [1977] suggested was used where the size strata were defined so that they were equal in terms of the square root of the size measure (in this case, the number of employees). The square root was used as a compromise between the needs of mine-level estimation where equal selection probabilities were best (size = 1) and employee-level estimation was best (size = number of employees). Detailed sample size allocation tables based on 2002 data for coal, metal, nonmetal, stone, and sand and gravel mines can be found in Appendix F.

Following the Office of Management and Budget (OMB) approval to conduct the national survey, the final sampling allocations were updated using 2007 second quarter MSHA data. Nine sampling frames were constructed based on the mining sector and mine type (underground or surface). The sampling was conducted using the SurveySelect procedure in the SAS statistical software package (SAS Institute Inc., Cary, NC). Systematic random sampling within the employee size strata was used together with controlled sorting by the state where the mine was located. The latter was done to ensure that the sample of mines was geographically representative. All metal mines and all underground nonmetal mines were selected with certainty from every stratum. The final survey sample of mines consisted of 331 underground coal, 385 surface coal, 74 underground metal, 159 surface metal, 39 underground nonmetal, 286 surface nonmetal, 96 underground stone, 498 surface stone, and 453 sand and gravel, for a total of 2,321 mining operations. Tables 1–9 present the sample allocations by mining sector and mine type.

Table 1. Sample Allocation for Underground Coal Mines

Stratum	Number of Mines	Percentage of Total Mines	Number of Employees	Percentage of Total Employees	Sample Mines
1–9	146	25.4%	331	0.8%	56
10–25	118	20.5%	1,972	4.8%	68
26–50	117	20.3%	4,460	10.8%	67
51–75	58	10.1%	3,622	8.8%	35
76–100	32	5.6%	2,790	6.8%	22
101–250	61	10.6%	9,267	22.5%	49
251+	43	7.5%	18,750	45.5%	34
Total	**575**	**100%**	**41,192**	**100%**	**331**

Table 2. Sample Allocation for Surface Coal Mines

Stratum	Number of Mines	Percentage of Total Mines	Number of Employees	Percentage of Total Employees	Sample Mines
1–9	725	51.8%	2,147	5.7%	101
10–25	302	21.6%	4,945	13.1%	84
26–50	209	14.9%	7,305	19.4%	75
51–75	65	4.6%	4,057	10.8%	36
76–100	30	2.1%	2,612	6.9%	20
101–250	44	3.1%	7,235	19.2%	44
251+	25	1.8%	9,407	24.9%	25
Total	**1,400**	**100%**	**37,708**	**100%**	**385**

Table 3. Sample Allocation for Underground Metal Mines

Stratum	Number of Mines	Percentage of Total Mines	Number of Employees	Percentage of Total Employees	Sample Mines
1–9	26	35.1%	137	1.8%	26
10–25	8	10.8%	134	1.8%	8
26–50	9	12.2%	327	4.3%	9
51–75	7	9.5%	443	5.8%	7
76–100	2	2.7%	168	2.2%	2
101–250	13	17.6%	2,312	30.4%	13
251+	9	12.2%	4,077	53.7%	9
Total	**74**	**100%**	**7,598**	**100%**	**74**

Table 4. Sample Allocation for Surface Metal Mines

Stratum	Number of Mines	Percentage of Total Mines	Number of Employees	Percentage of Total Employees	Sample Mines
1–9	60	37.7%	217	0.9%	60
10–25	21	13.2%	325	1.3%	21
26–50	13	8.2%	454	1.9%	13
51–75	4	2.5%	239	1.0%	4
76–100	3	1.9%	254	1.0%	3
101–250	26	16.4%	4,518	18.7%	26
251+	32	20.1%	18,204	75.2%	32
Total	**159**	**100%**	**24,211**	**100%**	**159**

Table 5. Sample Allocation for Underground Nonmetal Mines

Stratum	Number of Mines	Percentage of Total Mines	Number of Employees	Percentage of Total Employees	Sample Mines
1–9	7	17.9%	30	0.6%	7
10–25	5	12.8%	95	1.9%	5
26–50	6	15.4%	258	5.2%	6
51–75	4	10.3%	257	5.2%	4
76–100	2	5.1%	170	3.5%	2
101–250	11	28.2%	1,980	40.3%	11
251+	4	10.3%	2,125	43.2%	4
Total	**39**	**100%**	**4,915**	**100%**	**39**

Table 6. Sample Allocation for Surface Nonmetal Mines

Stratum	Number of Mines	Percentage of Total Mines	Number of Employees	Percentage of Total Employees	Sample Mines
1–9	339	53.6%	1,305	7.5%	92
10–25	128	20.2%	1,990	11.4%	65
26–50	81	12.8%	3,052	17.6%	46
51–75	32	5.1%	2,026	11.7%	32
76–100	19	3.0%	1,689	9.7%	17
101–250	25	3.9%	3,805	21.9%	25
251+	9	1.4%	3,520	20.2%	9
Total	**633**	**100%**	**17,387**	**100%**	**286**

Table 7. Sample Allocation for Underground Stone Mines

Stratum	Number of Mines	Percentage of Total Mines	Number of Employees	Percentage of Total Employees	Sample Mines
1–9	15	14.0%	78	2.0%	14
10–25	42	39.3%	733	18.8%	32
26–50	30	28.0%	1,030	26.4%	30
51–75	13	12.1%	812	20.8%	13
76–100	2	1.9%	174	4.5%	2
101–250	4	3.7%	511	13.1%	4
251+	1	0.9%	560	14.4%	1
Total	**107**	**100%**	**3,898**	**100%**	**96**

Table 8. Sample Allocation for Surface Stone Mines

Stratum	Number of Mines	Percentage of Total Mines	Number of Employees	Percentage of Total Employees	Sample Mines
1–9	2,034	49.6%	9,039	11.9%	116
10–25	1,345	32.8%	21,224	28.0%	114
26–50	426	10.4%	15,002	19.8%	95
51–75	107	2.6%	6,537	8.6%	51
76–100	56	1.4%	4,903	6.5%	36
101–250	128	3.1%	18,294	24.1%	83
251+	3	0.1%	911	1.2%	3
Total	**4,099**	**100%**	**75,910**	**100%**	**498**

Table 9. Sample Allocation for Sand and Gravel Mines

Stratum	Number of Mines	Percentage of Total Mines	Number of Employees	Percentage of Total Employees	Sample Mines
1–3	2,846	44.3%	5,555	13.0%	119
4–6	1,615	25.1%	7,761	18.2%	80
7–9	729	11.3%	5,682	13.3%	37
10–25	1,010	15.7%	14,629	34.4%	110
26–50	190	3.0%	6,411	15.1%	70
51–75	27	0.4%	1,632	3.8%	27
76–100	8	0.1%	684	1.6%	8
101–250	2	0.0%	219	0.5%	2
251+	0	0.0%	0	0.0%	0
Total	**6,427**	**100%**	**42,573**	**100%**	**453**

Data Collection

Survey Packet

The survey packet mailed to each sampled mining operation contained the following materials:

- A cover letter from NIOSH that introduced the study to the selected mines and stressed the importance of the study to the safety and health of miners. The letter was personalized and addressed to the best respondent identified through initial contacts with the mine.
- A Questions and Answers brochure that answered frequently asked questions.
- A copy of the paper questionnaire.
- Personalized directions for accessing the Internet questionnaire.
- A postage-paid return envelope for returning the hard-copy questionnaire.

The mine respondents were given the option of completing either the paper questionnaire booklet or the web-based survey questionnaire. The Questions and Answers brochure explained that both surveys asked the same questions. To minimize the employee-level questionnaire burden, mines with 30 or more employees were asked to provide data for only a sample of the total employees working during the specific reporting week. Mines with less than 30 employees were asked to report for all of them.

For mines with 30 or more employees working in the reference week, the mine respondent was asked to select the sample of employees by following sampling instructions included in the survey questionnaire. The sampling instructions were designed to select from 15 to 25 employees per mine, with employee counts from the frame used to determine the sampling rate. The employees were selected using systematic sampling with custom-generated "start-with" and "take-every" numbers included on the instructions page of the questionnaire. The "take-every" number was determined by dividing the number of employees the mine reported to MSHA by 30 and then rounding down. A random number table was consulted to get a random number between 1 and the "take-every" number which would be the "start-with" number. The "start-with" number constituted the first selection made from the list, prepared by the mine, of employees working during the reference week. The "take-every" number needed to be added repeatedly to the "start-with" number to determine the remaining selections. The variable number of employees selected per mine resulted from the need to use an integer as the "take-every" number to simplify the mathematics for the respondent.

The MSHA employment data printed on the mine's questionnaire may not have been current for the data collection period. This limitation was handled by instructing the mine respondents to call in when their mine employment for the reference week was 20 percent greater or 20 percent less than the employment projected from the MSHA data. The survey contractor would then provide alternative "start-with" and "take-every" numbers to these mine respondents, after determining that the respondents were reporting for the correct mine.

Each sampled mining operation was randomly assigned a reporting week, balanced by mine type and sector. The reporting week was a seven-day period that the mine respondent was asked to reference when answering some items in the questionnaire. The reporting week was described in the questionnaire as the mine's payroll week, which included the date that was preprinted on the first page of the questionnaire. Over the course of the survey, there were a total of 12 reporting weeks. On average, 193 mines were assigned to each reporting week (see Table 10).

Table 10. Number of Mines in the Final Sample by Sector, Type, and Reporting Week

Week	Total Mines	Coal	Metal	Nonmetal	Stone	Sand and Gravel	Surface	Underground
1	193	59	20	27	50	37	147	46
2	193	60	19	27	49	38	148	45
3	194	60	19	28	50	37	148	46
4	193	60	19	27	49	38	148	45
5	194	60	19	27	50	38	149	45
6	194	60	19	27	50	38	149	45
7	193	59	20	27	49	38	149	44
8	194	60	20	27	50	37	148	46
9	193	60	19	27	49	38	148	45
10	193	59	20	27	49	38	149	44
11	193	59	20	27	49	38	149	44
12	194	60	19	27	50	38	149	45
Total	2,321	716	233	325	594	453	1,781	540

Survey Promotion

Several initiatives were implemented before the start of data collection to promote the survey and to maximize response rates. OMSHR undertook considerable efforts to publicize the survey. At the start of data collection in March 2008, the National Mining Association offered to prepare and publish an article about the survey in its newsletter. Throughout the data collection period, OMSHR continued to pursue additional publicity efforts, promoting the survey both within NIOSH and to the mining community. A sand and gravel industry newsletter included an article about the survey. A notice about the survey was also published in the May 2008 issue of *CoalUSA* magazine.

Prior to the mailing of the survey packet, initial telephone calls were made to the contacts identified for each selected mine. In some cases the same contact individuals were found to be associated with multiple mines; for example, nine contacts were affiliated with mining companies that each had seven or more mines in the sample. A special effort was made by OMSHR to contact these individuals and inform them of the selection of multiple mining operations, determine the most appropriate addressee/recipient of the survey packet, and encourage participation in the survey. Throughout the data collection period, OMSHR continued to assist the survey contractor in both initiating and receiving calls with mine contacts and in responding to e-mails from the sampled mining operations.

Follow-up Contacts

Once the survey packet had been sent to the contact person at the mine, the data collection schedule provided for a three-week waiting period, to allow the contact the opportunity to complete the survey. After the waiting period, follow-up reminder calls were made to those mines that did not return their questionnaires or complete the web surveys by the "please submit" date printed on the survey. The main functions of the follow-up calls were to:

- Ensure that the survey materials had been received and that the materials were delivered to the appropriate person.
- Answer any questions regarding completing the survey.
- Serve as a reminder to complete the survey.

The most difficult challenge of follow-up was simply reaching the contacts. To deal with this issue, various approaches and initiatives were implemented. Because e-mail addresses were often available for mine contacts, an e-mail initiative was developed whereby an e-mail reminder was sent to anyone who had: (1) started, but did not complete a web survey; (2) not yet opened a web survey; (3) not returned a questionnaire; or (4) not made contact during the follow-up calls. This resulted in some immediate responses to the e-mails, along with many calls to the toll-free study telephone line and directly to OMSHR, often from contact persons who had a question on how to complete the survey. There were also a number of out-of-office replies that were useful in determining when another follow-up attempt could be made.

In addition, OMSHR also prepared a follow-up letter, cosigned by the study project director and the director of the NIOSH Office of Mine Safety and Health Research, with space at the bottom for the web survey login information and mine-specific password. This letter was mailed to contacts at more than 1,000 mining operations. As a result, OMSHR received some additional completed questionnaires. However, a large number of letters were returned as undeliverable.

Data Imputation and Statistical Weighting Procedures

A questionnaire was considered completed if it was missing no more than two of the 52 critical items listed in Appendix G. Returned questionnaires with more than two missing critical items were considered partially complete and, when possible, data imputation was used to complete these missing items.

Data Imputation

Imputation is the process of replacing missing data with legitimate values derived through logical deduction, regression models, or other probabilistic means. For the National Survey of the Mining Population, an attempt was made to impute missing data for the questions in the Training; Prep Plant/Mill Workers (found in the Work Schedules and Shift Work sections); Independent Contractor Employees; Safety, Communication, and Rescue Measures; and Employee Length of Service sections of the questionnaire. The Prep Plant/Mill Workers questions were imputed via information retrieved from the MSHA data on mines not having a preparation plant. In these cases, the relevant questionnaire items were set equal to zero or to the "not applicable" response. The questions in the Training; Independent Contractor Employees; and Safety, Communication and Rescue Measures sections were imputed via logical deduction, that is, when one or more responses were affirmative within the section and no negative responses were recorded, all missing items were set to the negative response. The Employee Length of Service section was completed via a regression model that predicted one or more missing items for the Total Years in this Job Title, Total Years at this Mine, and Total Years in Mining questions from those of the three that were reported.

Data Weighting, Estimation, and Variance Estimation

Sample survey data are weighted in order to provide unbiased or nearly unbiased estimates. The weights take into account variable probabilities of selection as well as compensate for bias introduced by differences between respondents and nonrespondents. For the National Survey of the Mining Population, weights were calculated in two steps. First, a base weight was calculated as the reciprocal of a given mine's probability of selection. These probabilities varied by major mining sector (coal, metal, nonmetal, stone, and sand and gravel), mine type (underground or surface), and mine size (number of employees). Second, a nonresponse adjusted weight was calculated as the product of the base weight and a nonresponse adjustment factor. The nonresponse adjustment factor was calculated as the ratio of the sum of weights for all eligible mines within a primary stratum (sector by mine type) to the sum of the weights for all responding mines.

Survey sampling implies some imprecision in the estimates and this imprecision is measured as variance and standard errors. For this survey, the Jackknife Repeated Replication (JRR) method was used to support variance estimation. One hundred replicate weights were created for each record in the dataset, with every replicate weight repeating the two steps described previously. Each replicate weight was used to derive a replicate estimate, and the variance in the replicate estimates (across the 100 replicates) could then be used to estimate the variance and standard error of each survey estimate.

Lessons Learned

The following lessons learned are based on project staff observations (by both OMSHR and the survey contractor) and the feedback obtained from the survey respondents. Also presented are any additional methods that could have been implemented to potentially increase the response rate or the efficiency of the study management.

- The questionnaire, with its foldout employee section and sample selection approach, appeared to be a barrier to completion. It is possible that the perceived level of effort for completing the survey prevented some mines from participating.
- Comments regarding ease of use of the survey were similar for both paper and web-based respondents. Partial responses on both versions of the questionnaire often stopped at the beginning of the employee section. This may have occurred when the mine contact realized that he/she could not complete the full questionnaire without retrieving information from other people in the mine organization, or from records not conveniently

located, or that other staff may have been unable or unwilling to complete the questionnaire.
- There was no incentive provided for completion of the survey, other than to assist OMSHR.
- Most refusal information related to time/burden issues. Some contacts refused after learning that they were assigned to complete questionnaires for multiple mines in their organization. Health and safety contacts often said that the survey content focused too little on health and safety issues and too much on human resource questions.
- The e-mail follow-up reminders and OMSHR follow-up letter were helpful initiatives, but may have been more effective had they been initiated at an earlier time in the data collection process.
- Even though multiple contact attempts and various response modes were used in this survey, conducting a nonresponse survey could have helped to ascertain whether the population of nonresponders differed measurably from the participants. It also could have been very useful in understanding and characterizing barriers to participation.
- Some suggestions for future surveys are the following:
  - Conducting the survey to focus on one major mining sector at a time in order to improve performance.
  - Reducing the length of the questionnaire, in response to complaints from mines that did not have the staff or the time to complete it.
  - Involving large mining companies in early reviews of the survey to obtain their input on questions they might find objectionable and their feedback on how to best administer the survey.

Survey Results

Overall, 954 completed or partial surveys were returned from the sampled mining operations. The outcomes of data collection for each of the sampled mines are summarized in Table 11. The 651 "critical items complete" and 86 "final missing critical items" questionnaires were the 737 survey responses that were used for the estimates presented in this IC. The mode of completion by the respondents is shown in Table 12.

Table 11. Summary of Final Results for All Sampled Mines

Result Code Description	Total
Critical items complete *	651
Final missing critical items *	86
Partial response	217
Final refusal—explicit refusal by corporate management	56
Final refusal, other reason—explicit refusal by local mine management	77
Final refusal, records unavailable—explicit refusal by local mine management	5
Final refusal, staff time—explicit refusal by local mine management	85
Ineligible mine	85
Ineligible, no contact	32
Initialized, no response	1,020
Hard-copy questionnaire received, but blank	7
Total	**2,321**

*Comprised final survey dataset

Table 12. Number of Completed Surveys by Mode

Mode	Count	Percentage
Web questionnaire	360	49%
Paper questionnaire	377	51%
Total	**737**	**100%**

Based on the review of the results of all contact attempts, 117 mines were determined to be ineligible. A summary of the ineligible mines by sector is presented in Table 13. Some of the reasons for ineligibility were:

- Mine has been closed.
- No contact was ever made with anyone at the mine.
- Mine is nonproducing.
- Construction work on the mine has not yet begun.
- Mine is shutting down and moving out equipment.
- Mine was just an exploration mine and was never in a producing status.
- Mine contracts out all of its mining operations.

Table 13. Summary of Ineligible Mines by Sector

Mine Sector	Ineligible Mine	Ineligible No Contact	Total
Coal	53	14	67
Metal	10	7	17
Nonmetal	9	4	13
Stone	7	3	10
Sand and Gravel	6	4	10
Total	**85**	**32**	**117**

Refusals to participate in the survey were received from 223 mines. The major reasons for refusal are shown in Table 14.

Table 14. Summary of Refusal by Mine Sector and Type of Refusal

Reason for Refusal	Coal	Metal	Nonmetal	Stone	Sand and Gravel	Total
Corporate refusal	20	1	11	16	8	56
General refusal	18	7	13	23	16	77
Records unavailable	1	3	0	0	1	5
Staff time	27	8	12	16	22	85
Total	**66**	**19**	**36**	**55**	**47**	**223**

The overall weighted response rate for the survey was 36.7 percent, with the lowest response rate for coal mines (25.8 percent) and the highest for nonmetal mines (48.8 percent). Underground mines responded at 30.1 percent compared to surface mines at 37.1 percent. The response rate data are presented in Figure 9.

The weighted response rates were calculated as the ratio of the sum of the weights of responding mines divided by the sum of the weights for all eligible sampled mines. The denominator included all nonresponding mines that were known to be eligible along with a percentage, p, of the weight corresponding to mines which did not respond but for whom it was not possible to determine whether in fact they were eligible. The percentage, p, was computed as the ratio of the weights of known eligible nonrespondents, plus respondents, plus ineligible mines. The ratios were computed separately for each nonresponse adjustment cell, which was defined by sector, mine type, and mine size.

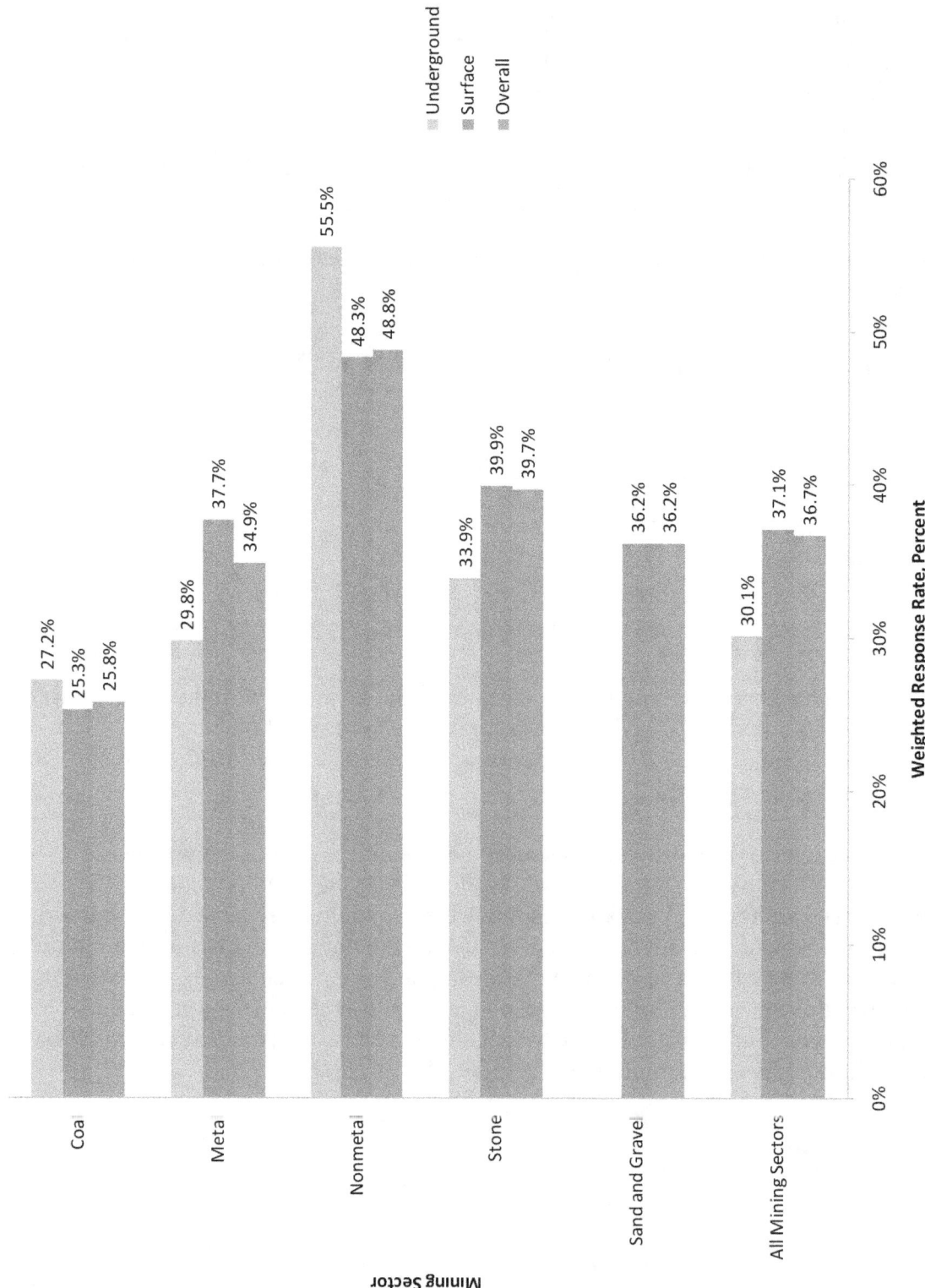

Figure 9. Weighted Response Rates by Sector and Mine Type.

Based on the data collected in this survey, Table 15 represents national estimates of the number of mines and the mine operator employees (with associated 95 percent confidence intervals) by sector during a typical week in the spring/summer of 2008. There were an estimated 231,549 employees working in 12,321 mines. Of these employees, 53,326 worked in 668 underground mines and the remaining 178,222 worked in 11,654 surface mines.

Table 15. National Estimates of Mines and Mine Employees in Spring/Summer 2008

Mine Sector	Number of Mines*	95% CI	Number of Employees*	95% CI
Coal, underground	454	(411, 498)	38,290	(31,088, 45,492)
Coal, surface	1,053	(925, 1,181)	31,717	(23,810, 39,625)
Metal, underground	71	(62, 79)	8,653	(2,419, 14,887)
Metal, surface	130	(113, 147)	30,430	(9,332, 51,528)
Nonmetal, underground	38	(29, 47)	3,424	(1,919, 4,928)
Nonmetal, surface	577	(506, 647)	15,925	(10,668, 21,182)
Stone, underground	105	(92, 118)	2,959	(2,491, 3,427)
Stone, surface	3,852	(3,600, 4,104)	68,006	(62,641, 73,372)
Sand and Gravel	6,042	(4,774, 7,309)	32,144	(26,275, 38,013)
Total	12,321	(11,003, 13,640)	231,549	(204,685, 258,413)

*Data do not sum to total due to independent rounding.

Employee Job Titles

The information for the mine operator employee job titles was collected using an open-ended format in which the survey respondent was asked to "write in" the job title for each of his/her sampled employees. A detailed listing of the job titles supplied by the respondents can be found in Appendix H. This approach allowed flexibility and lessened burden by not constraining the respondent to determine the most appropriate fit from a list of predefined job-title categories.

Initially, the Mine Safety and Health Administration (MSHA) Part 50 Data User's Handbook [MSHA 2007] was used to code the job titles supplied by the survey respondents. In some cases, slang terms or the name of a piece of mining equipment were provided as the employee's job title. To handle these situations, job codes were assigned by researching the equipment, mine type or commodity, and consulting *A Dictionary of Mining, Mineral, and Related Terms* [Thrush 1968] and *The Dictionary of Mining and Mineral Terms* [Infomine Inc. 2010]. Mining program researchers who were former mine employees also assisted by reviewing the "difficult-to-code" job titles and defining the occupation. In some instances, where logical, multiple job titles were combined under a single occupation code. For example, a "Belt Worker" and "Belt Man" were assigned the same code.

Once the job titles were coded, they were grouped into occupational categories. The four major categories are Administration/Professional, Maintenance, Production, and Service and

Utility. When a reported job title did not fit within any of these four categories, it was put into a Miscellaneous category. Within the four major occupational categories, there are subunits with up to four levels. Each of these subunits is further divided based on the type of work performed. National estimates of the number of workers have been computed for each major category (excluding Miscellaneous where only survey counts are reported) and the first three subunit levels.

Statistical Analysis

The statistical analysis of the data from the National Survey of the Mining Population was conducted using the SAS statistical software package. The SAS SURVEYFREQ and SURVEYMEANS procedures were used to create the weighted summary statistics that are reported in the IC. These procedures properly analyze data from complex sample surveys by taking the sample design into account. The variance estimation method used with these data was the Jackknife Repeated Replication (JRR). At this time, the subpopulation analysis for JRR is not available in SAS 9.2. In order to provide national estimates for the coal, metal, nonmetal, stone, and sand and gravel mining sectors, a SAS macro, using a reweighting method, proposed by Wang and Waldron [2010] was adopted for these subpopulation analyses. In their paper, Wang and Waldron compared the results of a subgroup analysis using their macro with PROC SURVEYMEANS and found these results were almost identical to those obtained when using the standard subpopulation analysis procedure in both the Stata 10.0 (StataCorp LP) and SUDAAN 10 (RTI International) statistical analysis software packages. In order to provide a measure of precision, a 95 percent confidence interval (CI) has been calculated for all survey estimates reported in this IC. Data were suppressed, and no national estimates were computed when the unweighted survey count was fewer than five responses (i.e., the number of responses was too small to be able to produce a reliable estimate) [NCHS 2002, 2004]. Due to independent rounding, the percentages shown in the individual bar charts may not sum to exactly 100 percent.

Employee Statistics for All Mines

Summary of Employee Statistics for All Mines

The demographic and occupational characteristics of employees in the U.S. mining industry are presented in Tables 16 and 17 and Figures 10–12. The weighted estimate for gender indicates that the workforce is composed predominately of male employees (92.5 percent). The major racial category is White (93.6 percent) followed by Black or African American (4.3 percent). Twelve percent of these employees have an ethnicity of Hispanic or Latino. Sixty-five percent are high school graduates, with another 24.1 percent having an education level beyond high school. A review of the weighted estimates indicates that the average worker is 43.3 years of age and has worked in mining for 12.9 years, 9.0 years at the current mine, and 7.1 years in his/her job title. The number of hours worked per week averages 45.4 with the "Surface Mine: Strip, Open Pit or Quarry" being the primary work location for the majority, or 34.1 percent of miners. An additional 23.0 percent of employees work in "Mill Operations, Preparation Plants, or Breakers," and another 18.3 percent are employed in the "Underground Mine: Underground" work location.

Tables 18, 19, 21, 22, and Figure 13 present the national estimates of the number of workers by four major occupational categories. (No estimates were calculated for Table 20: "Miscellaneous.") An estimated 62,646 (27.2 percent) mine workers are employed in the "Administration/Professional" category; 35,276 (15.3 percent) in the "Maintenance" category; 90,495 (39.4 percent) in the "Production" category; and 41,851 (18.2 percent) in the "Service and Utility" category.

Table 16. Demographic Characteristics of Employees at All Mines

Demographic Characteristic	Survey Count	National Estimate	95% LCL	95% UCL	National Percent	95% LCL	95% UCL
Gender:							
Male	8,414	211,471	188,671	234,270	92.5	91.1	93.9
Female	577	17,213	12,403	22,024	7.5	6.1	8.9
Age (years)	8,673	43.3	42.4	44.1			
Highest level of education:							
Less than 9th grade	222	4,996	3,062	6,930	2.4	1.5	3.3
9th–12th grade (no diploma)	800	18,600	15,299	21,902	8.8	7.3	10.3
HS Graduate or Equivalent (GED)	5,452	136,599	121,769	151,429	64.7	61.3	68.1
Some College, Associate Degree, or Technical School	1,392	39,326	30,655	47,996	18.6	15.9	21.3
Bachelor's Degree or beyond	452	11,516	9,017	14,014	5.5	4.5	6.4
Ethnicity:							
Hispanic or Latino	927	26,622	17,120	36,123	12.1	8.9	15.4
Non-Hispanic or Non-Latino	7,766	192,839	172,663	213,016	87.9	84.6	91.1
Race:							
American Indian or Alaska Native	119	4,050	1,851	6,249	2.0	0.9	3.0
Asian	9	183	56	311	0.1	0.0	0.2
Black or African American	397	8,893	6,419	11,367	4.3	3.2	5.4
Native Hawaiian or Other Pacific Islander	14	634	140	1,127	0.3	0.1	0.5
White	7,717	194,016	174,955	213,077	93.6	92.1	95.0

Table 17. Occupational Characteristics of Employees at All Mines

Occupational Characteristic	Survey Count	National Estimate	95% LCL	95% UCL	National Percent	95% LCL	95% UCL
Hours worked (per week)	8,363	45.4	44.6	46.2			
Experience:							
Experience in this Job Title (years)	8,641	7.1	6.4	7.8			
Experience at this Mine (years)	8,773	9.0	8.3	9.6			
Total Mining Experience (years)	8,539	12.9	12.1	13.7			
Primary Work Location:							
Underground Mine: Underground	1,585	42,191	34,049	50,333	18.3	15.0	21.6
Underground Mine: Surface Shops or Yards	287	4,884	3,461	6,307	2.1	1.5	2.7
Surface Mine: Strip, Open Pit, or Quarry	2,722	78,493	58,106	98,879	34.1	28.4	39.7
Surface Mine: Auger, Culm Bank, or Refuse Pile (Coal Mine Only)	78	3,581	267	6,896	1.6	0.1	3.0
Surface Mine: Dredge	185	4,491	2,551	6,430	1.9	1.1	2.8
Surface Mine: Other Surface Mining (Metal/Nonmetal Only)	922	21,492	14,757	28,227	9.3	6.4	12.3
Independent Shops or Yards	64	1,304	205	2,404	0.6	0.1	1.0
Mill Operations, Preparation Plants, or Breakers	2,251	53,052	45,563	60,541	23.0	19.2	26.9
Office	889	20,835	16,764	24,906	9.0	7.8	10.3

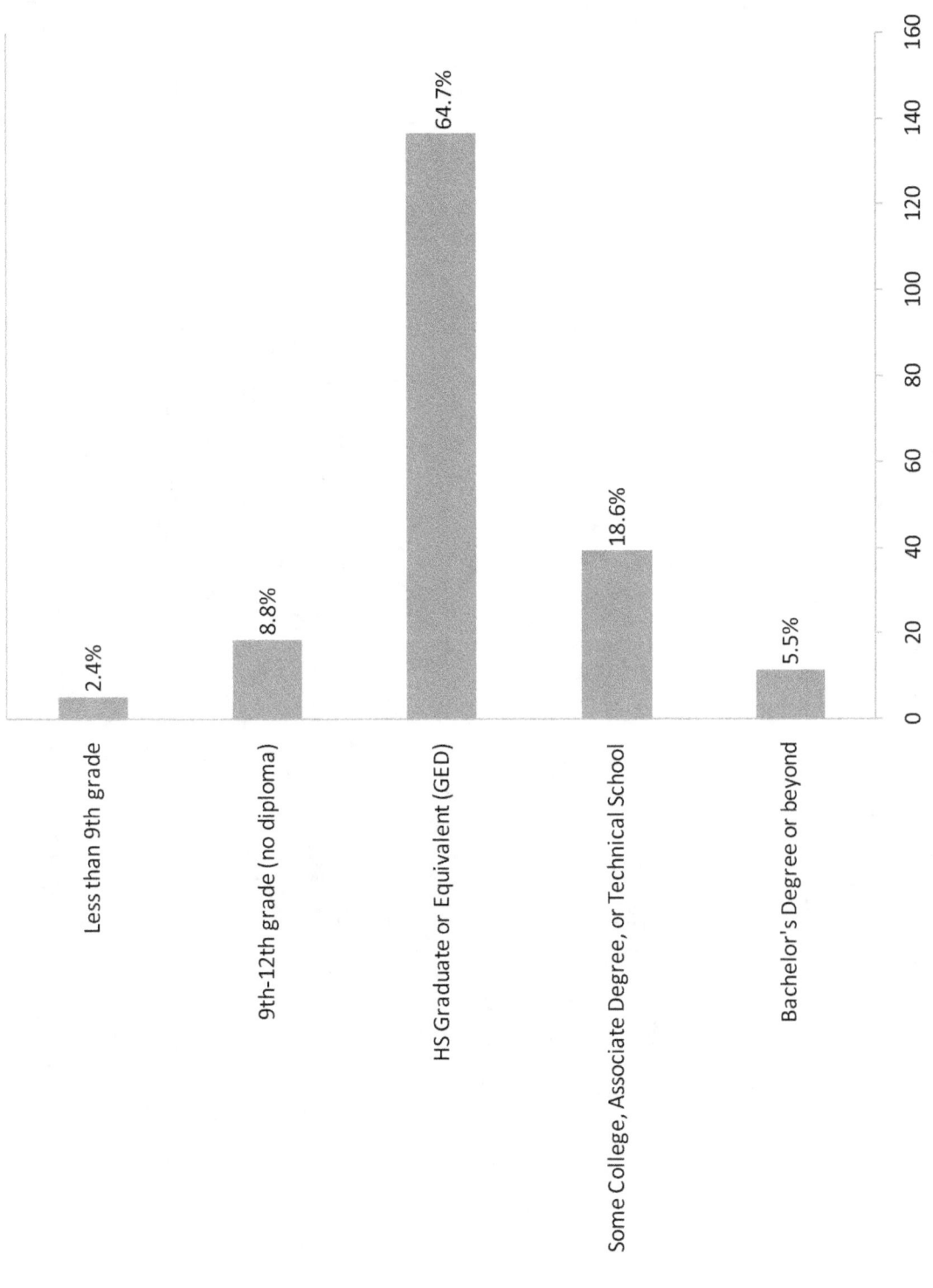

Figure 10. Education Level of Employees at All Mines.

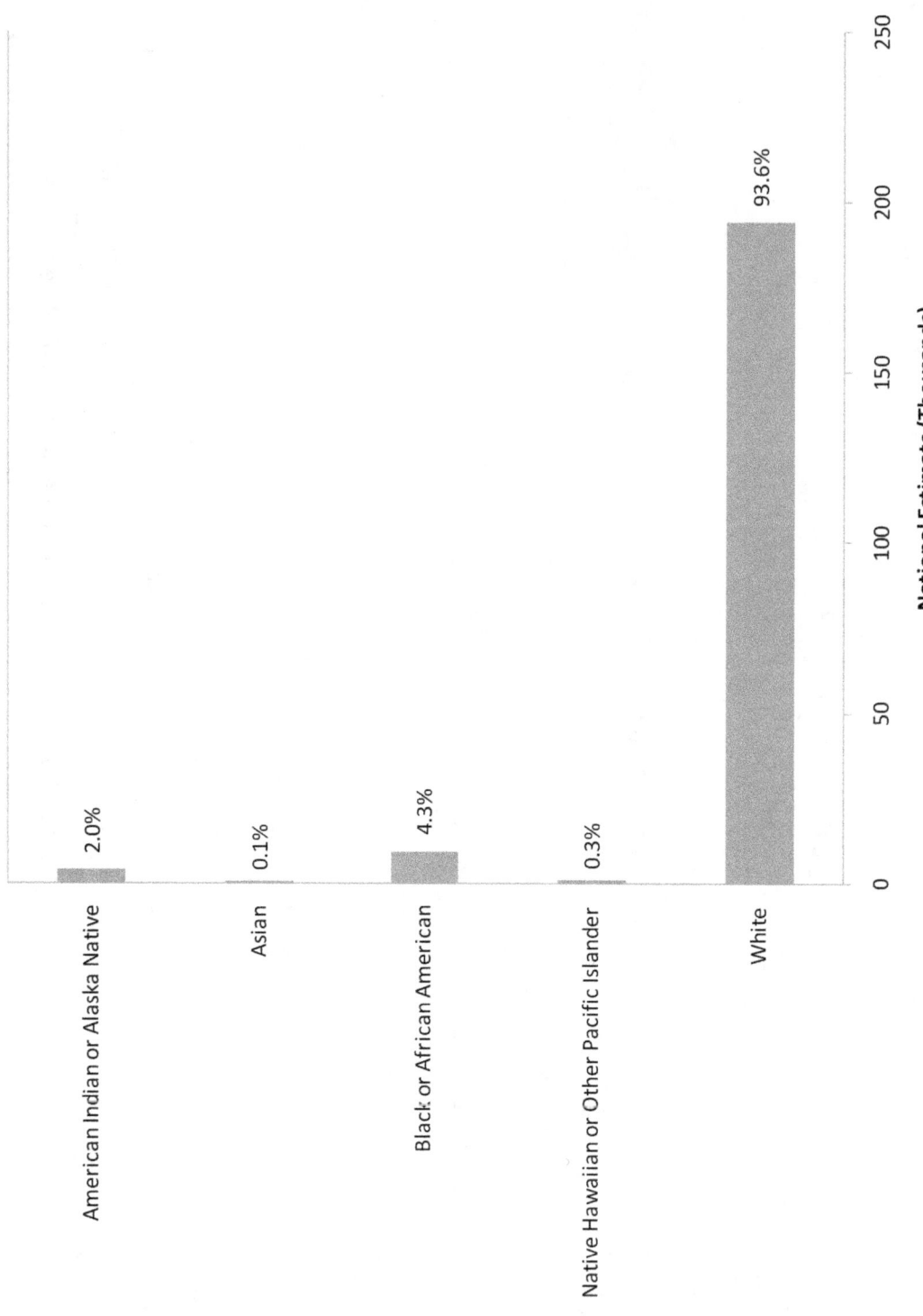

Figure 11. Race of Employees at All Mines.

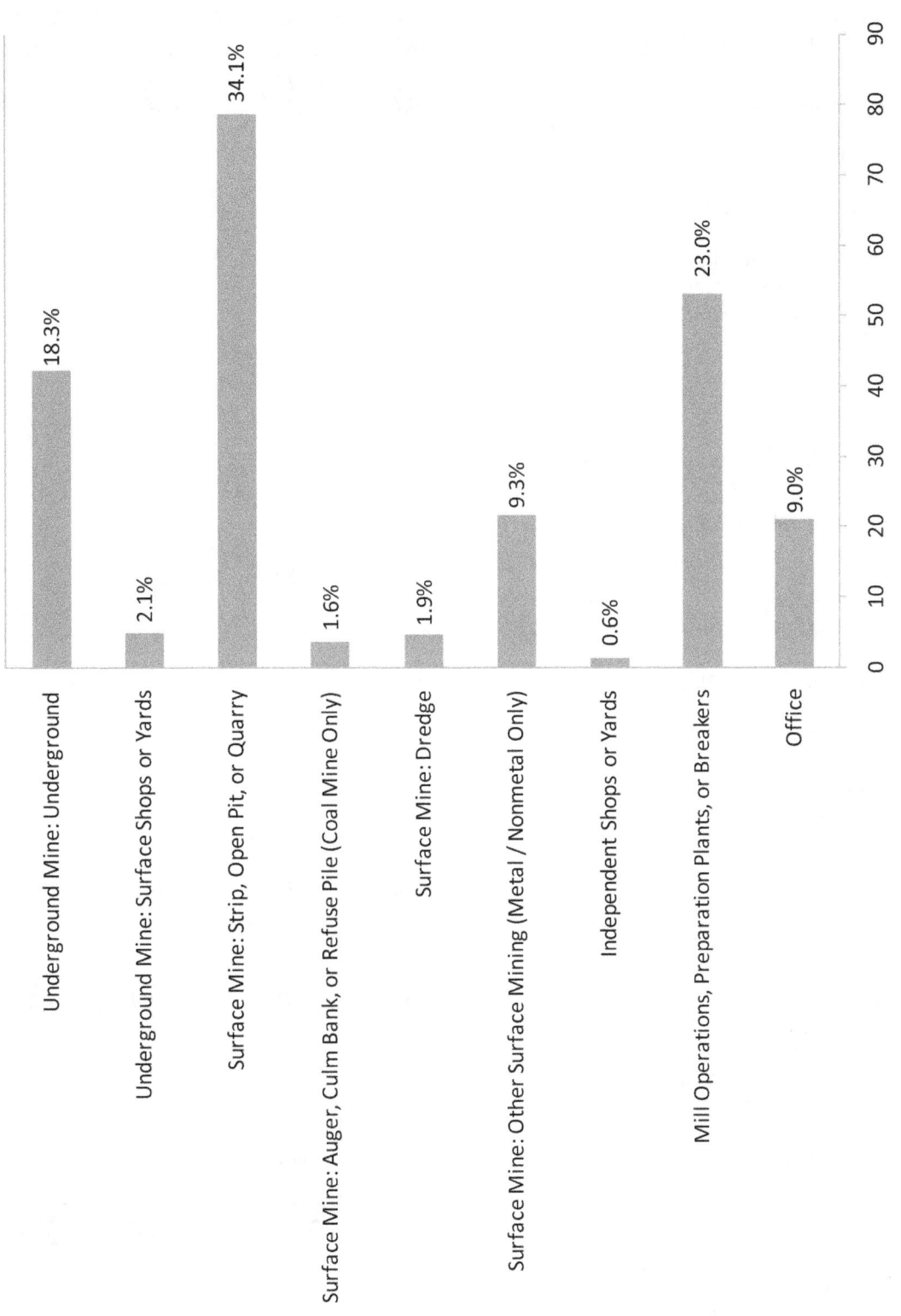

Figure 12. Primary Work Location of Employees at All Mines.

Table 18. Estimated Number of Administration/Professional Employees at All Mines

Occupation by Category	Survey Count	National Estimate	95% LCL	95% UCL
ADMINISTRATION/PROFESSIONAL	**2,453**	**62,646**	**54,584**	**70,708**
<u>Office Staff</u>	<u>408</u>	<u>10,181</u>	<u>8,304</u>	<u>12,057</u>
Administrative Staff	*225*	*5,493*	*4,299*	*6,687*
Administration				
Administrative Assistant				
Clerk				
Coal Distribution Coordinator				
Communications				
Customer Service				
Human Resources				
Information Technology				
Mine Clerk				
Office Clerk				
Office Staff				
Plant Clerk				
Receptionist				
Secretary				
Systems Analyst				
Technical Coordinator				
Business	*136*	*3,510*	*2,257*	*4,762*
Accounting				
Bookkeeper				
Buyer				
Cost Coordinator				
Payroll				
Procurement				
Purchasing				
Sales				
Shipping				
Terminal Operator				
Security	*14*	*346*	*119*	*573*
Guard				
Security				
Supplies	*31*	*818*	*404*	*1,232*
Supply Clerk				
Warehouse				
Warehouse Technician				
Warehouseman				
Union Representative	*2*	*DSU*	*DSU*	*DSU*
<u>Professional</u>	<u>334</u>	<u>10,304</u>	<u>7,332</u>	<u>13,276</u>
Engineer	*61*	*1,722*	*860*	*2,584*
Director of Engineering				
Engineer (Electrical/Mining/Ventilation)				
Engineer, not otherwise specified				
Environmental Engineer				
Plant Engineer				

Table 18. Estimated Number of Administration/Professional Employees at All Mines (continued)

Occupation by Category	Survey Count	National Estimate	95% LCL	95% UCL
Process Engineer				
Production Engineer				
Project Engineer				
Non-engineer	90	3,408	1,568	5,249
Chemist				
Control Person/Analyst				
Environmental				
Environmental Specialist				
Geologist				
Metallurgist				
Operating Engineer				
Operations				
Operations Specialist				
Physical Tester				
Planner				
Production Scheduler				
Reliability Engineer				
Surveyor				
Utility Engineer				
Technician	183	5,174	3,148	7,199
Coal Sampler				
Electrical Technician				
Electronic Technician				
Engineering Technician				
Fuel Operator/Technician				
Lab Technician				
Laboratory Technician/Refiner				
Materials Technician				
Mechanic Technician				
Mill Technician				
Mine Technician				
Plant Technician				
Process Control Operator/Technician				
Production Technician				
Quarry Technician				
Sampler/Lab Technician				
Technician				
Utility Technician				
Safety	**51**	**1,425**	**854**	**1,996**
Inspector				
Safety				
Safety Director				
Safety Manager				
Safety Supervisor				

Table 18. Estimated Number of Administration/Professional Employees at All Mines (continued)

Occupation by Category	Survey Count	National Estimate	95% LCL	95% UCL
<u>**Supervisory**</u>	<u>1,660</u>	<u>40,736</u>	<u>35,454</u>	<u>46,018</u>
Executive	*71*	*1,365*	*1,005*	*1,726*
CEO				
General Manager				
Mine Owner				
President				
Vice President				
Foreman	*626*	*15,807*	*12,870*	*18,744*
Assistant Superintendent				
Belt Foreman (underground)				
Electrical Foreman (underground)				
Foreman				
Foreman/Manager				
Foreman/Shift Boss				
Labor Foreman				
Lead Man				
Maintenance Foreman				
Maintenance Lead Man				
Mill Foreman				
Mine Foreman				
Outby Foreman				
Pit Foreman				
Plant Foreman				
Preparation Plant Foreman				
Production Foreman				
Section Foreman				
Section Foreman/Boss				
Shift Foreman				
Shop Foreman				
Superintendent				
Track Foreman				
Underground Foreman				
Manager	*339*	*8,224*	*6,266*	*10,182*
Area Manager				
Assistant Manager				
Assistant Mine Foreman/Assistant Mine Manager				
Concentrator Manager				
Customer Service Manager				
Distribution Manager				
Dredge Manager				
Dry Plant Manager				
Engineer/Operations Manager				
Engineering Manager				
Environmental Manager				
Equipment Maintenance Manager				
Equipment Manager				
Facility Manager				
Financial Manager				
Human Resources Manager				

Table 18. Estimated Number of Administration/Professional Employees at All Mines (continued)

Occupation by Category	Survey Count	National Estimate	95% LCL	95% UCL
Lab Manager				
Maintenance Manager				
Management				
Manager				
Mill Manager				
Mine Foreman/Mine Manager				
Mine Manager				
Office Manager				
Operations Manager				
Plant Manager				
Plant Superintendent				
Process Manager				
Production Manager				
Project Manager				
Purchasing Manager				
Quality Control Manager				
Quarry Manager				
Raw Material Manager				
Regulatory Manager				
Sales Manager				
Scale Office Manager				
Shift Manager				
Shipping Manager				
Shop Manager				
Storeroom Manager				
Technical Services Manager				
Supervisor	*624*	*15,340*	*13,052*	*17,627*
Assistant Mine Supervisor				
Assistant Supervisor				
Auger Crew Supervisor				
Backhoe Supervisor				
Bagging/Baghouse Supervisor				
Belt Coordinator				
Blasting Supervisor				
Clay Operator				
Concentrator Supervisor				
Control Room Supervisor				
Crusher Supervisor				
Dozer Supervisor				
Electrical Supervisor				
Engineering Supervisor				
Equipment Supervisor				
Gold House Supervisor				
Lab Supervisor				
Leaching Supervisor				
Loader Supervisor				
Loadhouse Supervisor				
Maintenance Supervisor				
Mechanic Supervisor				
Mine Operations				
Mine Operator				

Table 18. Estimated Number of Administration/Professional Employees at All Mines (continued)

Occupation by Category	Survey Count	National Estimate	95% LCL	95% UCL
Mine Supervisor				
Mobile Equipment Supervisor				
Pit Operator				
Pit Supervisor				
Plant Operator				
Plant Supervisor				
Prep Plant Operator				
Process Supervisor				
Production Supervisor				
Quality Assurance Supervisor				
Quarry Operator				
Quarry Supervisor				
Shift Supervisor				
Shipping Supervisor				
Supervisor				
Tailings Supervisor				
Transportation Supervisor				
Warehouse Supervisor				
Wash Plant Supervisor				

Abbreviation: DSU, data suppressed.

Table 19. Estimated Number of Maintenance Employees at All Mines

Occupation by Category	Survey Count	National Estimate	95% LCL	95% UCL
MAINTENANCE	**1,311**	**35,276**	**29,913**	**40,639**
<u>Specialty</u>	<u>272</u>	<u>8,234</u>	<u>6,445</u>	<u>10,022</u>
Electrician	*190*	*6,291*	*4,592*	*7,990*
Diagnostic Electrician				
Electrician				
Electrician/Wireman				
Electrician Trainee				
Maintenance Electrician				
Master Electrician				
Trainer Electrician				
Welder	82	1,942	1,312	2,572
Certified Welder				
Maintenance Welder				
Repair/Welder				
Welder				
Welder (nonshop)				
Welder/Fabricator				
Welder/Mechanic				
<u>Support</u>	<u>1,039</u>	<u>27,043</u>	<u>22,718</u>	<u>31,367</u>
Maintenance	*392*	*8,873*	*7,065*	*10,682*
Continuous Miner Maintenance				
Crusher Maintenance				
Dragline Oiler				
Electrical Maintenance				
Equipment Maintenance				
Fixed Maintenance				
Greaser/Oiler				
Liquid Fuel Handler				
Maintenance				
Maintenance Clerk				
Maintenance Coordinator				
Maintenance Planner				
Maintenance Technician				
Mechanic Clerk				
Mechanical Maintenance				
Mill Maintenance				
Millwright				
Mobile Maintenance				
Pipefitter				
Plant Maintenance				
Production/Process Maintenance				
Road Maintenance				
Skilled Maintenance				
Truck Maintenance				
Underground Belt Maintenance				
Underground Maintenance				

Table 19. Estimated Number of Maintenance Employees at All Mines (continued)

Occupation by Category	Survey Count	National Estimate	95% LCL	95% UCL
Mechanic	*556*	*14,368*	*11,607*	*17,129*
Aggregate Mechanic				
Automotive Mechanic				
Belt Mechanic				
Diagnostic Mechanic				
Diesel Mechanic				
Equipment Mechanic				
Heavy Equipment Mechanic				
Maintenance Mechanic				
Master Mechanic				
Mechanic				
Mechanic/Electrician				
Mechanic/Welder				
Mechanic Helper				
Mechanic Trainee				
Mobile Equipment Mechanic				
Mobile Maintenance Mechanic				
Mobile Mechanic				
Plant Mechanic				
Prep Plant Mechanic				
Shop Mechanic				
Underground Belt Mechanic				
Wrens Mechanic				
Repairman	91	3,801	739	6,864
Automotive Repairman				
Crusher Repairman				
Electronic/Electrical Repairman				
Heavy Duty Repairman				
Instrument Repairman				
Maintenance Repairman				
Mechanical Repairman				
Plant Repairman				
Repairman				
Skilled Repairman				
Tailings Repairman				
Underground Belt Repairman				
Underground Repairman				

Table 20. Number of Miscellaneous Employees at All Mines

Occupation by Category	Survey Count
MISCELLANEOUS	**35**
<u>Trainee</u>	<u>19</u>
<u>Unknown</u>	<u>16</u>

Table 21. Estimated Number of Production Employees at All Mines

Occupation by Category	Survey Count	National Estimate	95% LCL	95% UCL
PRODUCTION	**3,571**	**90,495**	**76,183**	**104,807**
<u>*Equipment Operator*</u>	<u>1,860</u>	<u>49,707</u>	<u>40,495</u>	<u>58,920</u>
Dragline Operator	24	677	275	1,079
Equipment Operator	944	23,373	19,276	27,469
Backhoe Operator				
Bobcat Operator				
Bulldozer Operator				
Crane Operator				
Dredge Operator				
End Dump Driver				
End Dump Driver/Operator				
Equipment Operator				
Forklift Operator				
Front End Loader				
Front End Loader Operator				
Grader Operator				
Gravity Mag Operator				
Heavy Equipment Operator				
Highlift Operator				
Hopper Operator				
Large Shovel/Backhoe/Load Operator				
Machine Operator				
Mobile Bridge Operator				
Mobile Equipment Operator				
Mucking Machine Operator				
Paver Operator				
Payloader Operator				
Raise Borer Operator				
Road Grader Operator				
Rock Duster				
Rotary Bucket Excavator Operator				
Rotary Dump Operator				
Scaler (mechanical)				
Scraper Operator				
Stationary Equipment Operator				
Stripping Operator				
Tower Operator				
Track Hoe				
Tractor Operator				
Tractor Operator/Motorman				
Hoist	36	430	93	766
Hoist Engineer				
Hoist Operator				
Hoistman				
Skip Tender/Cager/Station Attendant				

Table 21. Estimated Number of Production Employees at All Mines (continued)

Occupation by Category	Survey Count	National Estimate	95% LCL	95% UCL
Material Mover	704	18,923	14,005	23,841
Dump Truck Driver				
Haul Truck Operator				
Haul Truck Operator/Driver				
Hauler/Haul Unit Operator				
Hauler Operator				
Off Road Truck Driver				
Ore Truck Driver/Operator				
Pit Truck Driver				
Quarry Truck Driver				
Refuse Truck Driver/Backfill Truck Driver				
Rock Truck Driver				
Rubber Tire Operator				
Scoop Car Operator				
Scoop Loader				
Scoop Tram Operator				
Shuttle Car Operator				
Stock Truck/Stock Pile Driver				
Truck Driver				
Underground Coal Hauler				
Underground Haulage Operator				
Water Truck Operator				
Mining Machines	106	4,056	2,635	5,477
Continuous Miner Helper				
Continuous Miner Operator				
Face Operator				
Head Operator				
Jacksetter				
Longwall Operator				
Shearer Operator				
Undercutter Operator				
Operator/Driver	29	740	300	1,179
Dump Operator				
Motorman				
Motorman/Locomotive Operator				
Operator/Driver				
Transportation				
Shovel Operator	17	1,510	0	3,434
<u>Extraction Labor</u>	<u>212</u>	<u>5,229</u>	<u>2,590</u>	<u>7,868</u>
Coal Miner				
Heading Prep				
Mine Production				
Mine Spec				
Mine Support				
Miner				
Miner Support				
Production Miner				

Table 21. Estimated Number of Production Employees at All Mines (continued)

Occupation by Category	Survey Count	National Estimate	95% LCL	95% UCL
<u>**Material Preparation**</u>	<u>304</u>	<u>6,178</u>	<u>4,598</u>	<u>7,758</u>
Additives	14	271	9	532
Additive Press Operator				
Additives Utility				
Calcine Operator				
Thickener Operator				
Crusher	122	2,891	1,732	4,050
Blunging Operator				
Breaker Operator				
Crusher Attendant				
Crusher Helper				
Crusher Operator/Pan Feeder Operator				
Crusher Plant Operator				
Hammer Mill Operator				
Jaw Operator				
Mill Crusher Operator				
Rock Breaker Operator				
Screenhouse Crusher				
Cutter	70	1,194	286	2,102
Cutting Machine Operator				
Sawyer				
Splitter				
Stone Cutter				
Trimmer				
Mill	98	1,822	1,101	2,543
Dry Mill Operator				
Limestone Prep Operator				
Mill Hand/Helper				
Mill Operator (ball/pebble/rod)				
Mill Production Worker				
Milling Machine Operator				
Mill Man				
Roller Mill Operator				
Roller Operator				
<u>**Process**</u>	<u>186</u>	<u>5,769</u>	<u>2,649</u>	<u>8,890</u>
Belt Vulcanizer	9	464	0	1,197
Dry Processing	42	763	362	1,164
Dry Plant/Process Operator				
Dryer Operator				
Fluid Bed Dryer Operator				
Kiln Operator				
Other	32	825	300	1,351
Fabricator				
Process Attendant				
Process Operator				

Table 21. Estimated Number of Production Employees at All Mines (continued)

Occupation by Category	Survey Count	National Estimate	95% LCL	95% UCL
Separation	83	3,244	275	6,212
Centrifuge Utility				
Digestion Operator				
Extruder Operator				
Filter Evaporation Operator				
Filter Operator				
Flotation Plant Operator				
Flotation/Concentrator Operator				
Froth Cell Operator				
Grinder Operator				
Leach Utility				
Leaching Operations Worker				
Mix Chemist				
Mix Operator				
Pan Operator				
Pelletizing Operations Worker				
Pug Operator/Mixer Tender				
Rotex Operator				
Screen Plant Labor				
Screen Plant Operator				
Slurry Operator				
Tailings Operator				
Wash Process	12	410	105	715
Wash Operator				
Washer Operator				
Wet Processing	8	63	7	120
Wet Plant Attendant				
Wet Plant Operator				
Support	**1,009**	**23,611**	**19,052**	**28,170**
Drill Operator	101	1,684	1,179	2,189
Auger Operator				
Coal Drill Operator				
Drill Helper/Chuck Tender				
Drill Operator				
Highwall Drill Operator				
Rotary Electric/Hydraulic Drill Operator				
Electronics	4	DSU	DSU	DSU
Console Operator				
Power Systems				
Robot Operator				
Explosives	82	1,524	864	2,183
Blaster				
Driller/Blaster				
Explosives/Powder Man				
Shooter				
Shot Firer				

Table 21. Estimated Number of Production Employees at All Mines (continued)

Occupation by Category	Survey Count	National Estimate	95% LCL	95% UCL
Other	*579*	*14,279*	*9,944*	*18,614*
Control Room				
Controller				
Control Man				
Dispatcher				
Operator, not otherwise specified				
Panel Operator				
Port Operator				
Production Operator				
Rak Handler				
Scaler (hand)				
Top Operator				
Underground Operator				
Underground Plant Operator				
Quality Control	75	1,609	1,090	2,128
Quality Control				
Quality Control/Quality Assurance				
Roof Bolter	168	4,321	3,073	5,569
Roof Bolter				
Roof Control Operator				

Abbreviation: DSU, data suppressed.

Table 22. Estimated Number of Service and Utility Employees at All Mines

Occupation by Category	Survey Count	National Estimate	95% LCL	95% UCL
SERVICE and UTILITY	**1,638**	**41,851**	**36,921**	**46,781**
<u>General Labor</u>	<u>817</u>	<u>21,603</u>	<u>17,514</u>	<u>25,692</u>
Cleaners	*10*	*212*	*68*	*356*
Cleanup Man				
Dry Attendant				
Janitor				
Steamer				
Tank Car Washer				
Tower Cleaner				
Construction	*35*	*1,064*	*394*	*1,734*
Cement Man/Concrete Worker				
Construction				
Curb Cutter				
Ground Control/Timberman				
Packer				
Screed Person				
Shaft Miner/Shaft Repairer				
Laborer	*490*	*13,000*	*9,958*	*16,042*
Cook				
Ground Hand				
Ground Man				
Inside Laborer				
Laborer				
Miller				
Outby Laborer				
Outside Laborer				
Plant Helper				
Plant Man				
Production Support				
Production Worker				
Quarry Worker				
Root Picker				
Shop Man				
Stick Picker				
Surface Support				
Track Man				
Material Handling	*145*	*2,867*	*1,880*	*3,853*
Bagger/Bagging Operations Worker				
Crude Pile Operator				
Material Handler				
Mudpicker				
Palletizer				
Reclaim Operator				
Rolling Stock Crew				
Silo Operator				
Stacker				
Storage Operator				

Table 22. Estimated Number of Service and Utility Employees at All Mines (continued)

Occupation by Category	Survey Count	National Estimate	95% LCL	95% UCL
Storeroom				
Sweeper Operator				
Yard Laborer				
Yard Man				
Tradesman	33	1,928	0	3,966
Apprentice/Journeyman				
Boiler Operator				
Boilermaker				
Carpenter/Plumber/Painter				
Craftsman				
Machinist				
Trades Person				
Weighman	104	2,532	1,869	3,195
Scale Clerk/Operator				
Weighman				
Weighmaster				
Support Labor	**821**	**20,249**	**17,248**	**23,249**
Barge Operations	28	442	120	763
Barge Attendant/Boat Operator				
Boat Pilot				
Deck Hand				
Dock Hand				
Dock Worker				
Conveyor Operator	56	1,557	771	2,343
Belt Cleaner/Conveyor Man				
Belt Man/Conveyor Man				
Distribution	35	564	106	1,021
Packaging Operations Worker				
Packhouse				
Examiner	34	1,006	408	1,605
Fire Boss				
Mine Examiner				
Underground Belt Examiner				
Loading	462	11,020	9,083	12,958
Bin Attendant				
Bin Puller/Truck Loader				
Bulk Loader				
Chute Puller				
Load Haul Dump—Complete Cycle				
Loader Operator				
Loading				
Loadman				
Loadout Operator				
Operator/Loader				
Pit Loader Operator				
Plant Loader Operator				

Table 22. Estimated Number of Service and Utility Employees at All Mines (continued)

Occupation by Category	Survey Count	National Estimate	95% LCL	95% UCL
Production Loader				
Quarry Loader Operator				
Rail Loader Operator				
Shipping Loader				
Stock Loader/Piler				
Tipple Operator				
Underground Loader				
Yard Loader Operator				
Pumper	*10*	*467*	*0*	*982*
Gravel Pumper				
Pumper				
Supplies	*15*	*214*	*87*	*341*
Parts				
Parts Runner				
Supply Hauler				
Supply Man				
Supply Man/Nipper				
Utility	*177*	*4,853*	*3,328*	*6,377*
Crusher Utility				
E.O. Utility				
Equipment Utility				
Lampman				
Mill Utility				
Operator Utility				
Outside Utility				
Pit Utility Person				
Plant Utility				
Production Utility				
Quarry Utility				
Utility Beltline				
Utility Belts				
Utility Bolter				
Utility Lubricator				
Utility Man				
Utility Scaler				
Wet Utility				
Ventilation	*4*	*DSU*	*DSU*	*DSU*
Brattice Man				
Ventilation Man				

Abbreviation: DSU, data suppressed.

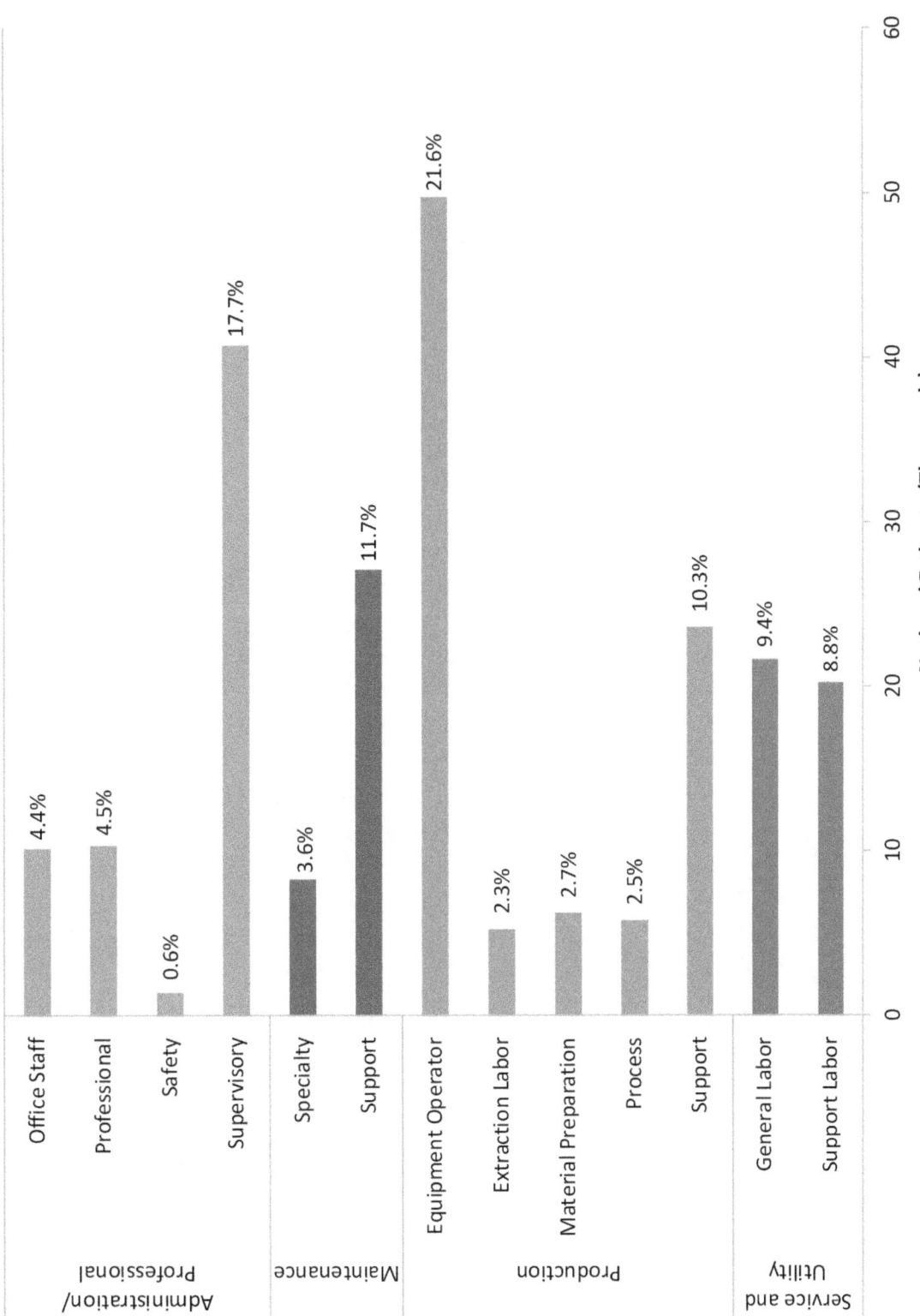

Figure 13. Occupational Categories of Employees at All Mines.

Employee Statistics for Coal Mines

Summary of Employee Statistics for Coal Mines

The demographic and occupational characteristics of employees in the U.S. coal mining industry are presented in Tables 23 and 24 and Figures 14–16. The weighted estimate for gender indicates that the workforce is composed predominately of male employees (96.2 percent). The majority of coal mine employees are White (96.4 percent) followed by American Indian or Alaska Native (2.5 percent). Only 1.9 percent of these employees have an ethnicity of Hispanic or Latino. Seventy-seven percent are high school graduates, with another 16.8 percent having an education level beyond high school. A review of the weighted estimates indicates that the average coal mine worker is 43.8 years of age and has worked in mining for 16.0 years, with 8.2 years at the current mine, and 7.8 years in his/her job title. The national estimate for the average number of hours worked per week is 47.3. The primary work location for an estimated 46.8 percent of coal mine employees is "Underground Mine: Underground." An additional 24.0 percent of these employees work at a "Surface Mine: Strip, Open Pit, or Quarry," while another 15.3 percent are employed in the "Mill Operations, Preparation Plants, or Breakers" work location.

Tables 25, 26, 28, 29, and Figure 17 present the national estimates of the number of coal mine workers by four major occupational categories. (No estimates were calculated for Table 27: "Miscellaneous.") An estimated 16,048 (23.2 percent) are employed in the "Administration/Professional" category; 12,000 (17.3 percent) in the "Maintenance" category; 29,562 (42.7 percent) in the "Production" category; and 11,791 (17.0 percent) in the "Service and Utility" category.

Table 23. Demographic Characteristics of Employees at Coal Mines

Demographic Characteristic	Survey Count	National Estimate	95% LCL	95% UCL	National Percent	95% LCL	95% UCL
Gender:							
Male	2,260	65,374	54,760	75,989	96.2	94.7	97.7
Female	66	2,559	1,406	3,713	3.8	2.3	5.3
Age (years)	2,255	43.8	42.5	45.1			
Highest level of education:							
Less than 9th grade	14	182	66	299	0.3	0.1	0.5
9th–12th grade (no diploma)	149	3,839	2,040	5,638	6.2	3.5	8.9
HS Graduate or Equivalent (GED)	1,644	47,548	38,760	56,336	76.7	72.4	80.9
Some College, Associate Degree, or Technical School	273	8,698	6,097	11,300	14.0	10.7	17.4
Bachelor's Degree or beyond	56	1,742	973	2,512	2.8	1.7	3.9
Ethnicity:							
Hispanic or Latino	37	1,222	430	2,015	1.9	0.7	3.0
Non-Hispanic or Non-Latino	2,224	64,548	53,859	75,237	98.1	97.0	99.3
Race:							
American Indian or Alaska Native	37	1,635	0	3,434	2.5	0.0	5.2
Asian	0	NA	NA	NA	NA	NA	NA
Black or African American	26	774	189	1,358	1.2	0.3	2.1
Native Hawaiian or Other Pacific Islander	0	NA	NA	NA	NA	NA	NA
White	2,209	62,528	51,932	73,125	96.4	93.5	99.3

Abbreviation: NA, not applicable.

Table 24. Occupational Characteristics of Employees at Coal Mines

Occupational Characteristic	Survey Count	National Estimate	95% LCL	95% UCL	National Percent	95% LCL	95% UCL
Hours worked (per week)	2,131	47.3	45.9	48.7			
Experience:							
Experience in this Job Title (years)	2,209	7.8	6.9	8.7			
Experience at this Mine (years)	2,294	8.2	6.6	9.8			
Total Mining Experience (years)	2,166	16.0	14.3	17.7			
Primary Work Location:							
Underground Mine: Underground	1,021	32,358	26,196	38,519	46.8	40.5	53.1
Underground Mine: Surface Shops or Yards	82	2,477	1,447	3,508	3.6	2.0	5.1
Surface Mine: Strip, Open Pit, or Quarry	613	16,620	11,106	22,135	24.0	17.5	30.6
Surface Mine: Auger, Culm Bank, or Refuse Pile (Coal Mine Only)	78	3,581	267	6,896	5.2	0.6	9.8
Independent Shops or Yards	19	462	0	1,344	0.7	0.0	1.9
Mill Operations, Preparation Plants, or Breakers	407	10,565	6,984	14,147	15.3	10.8	19.7
Office	107	3,103	1,956	4,249	4.5	3.0	6.0

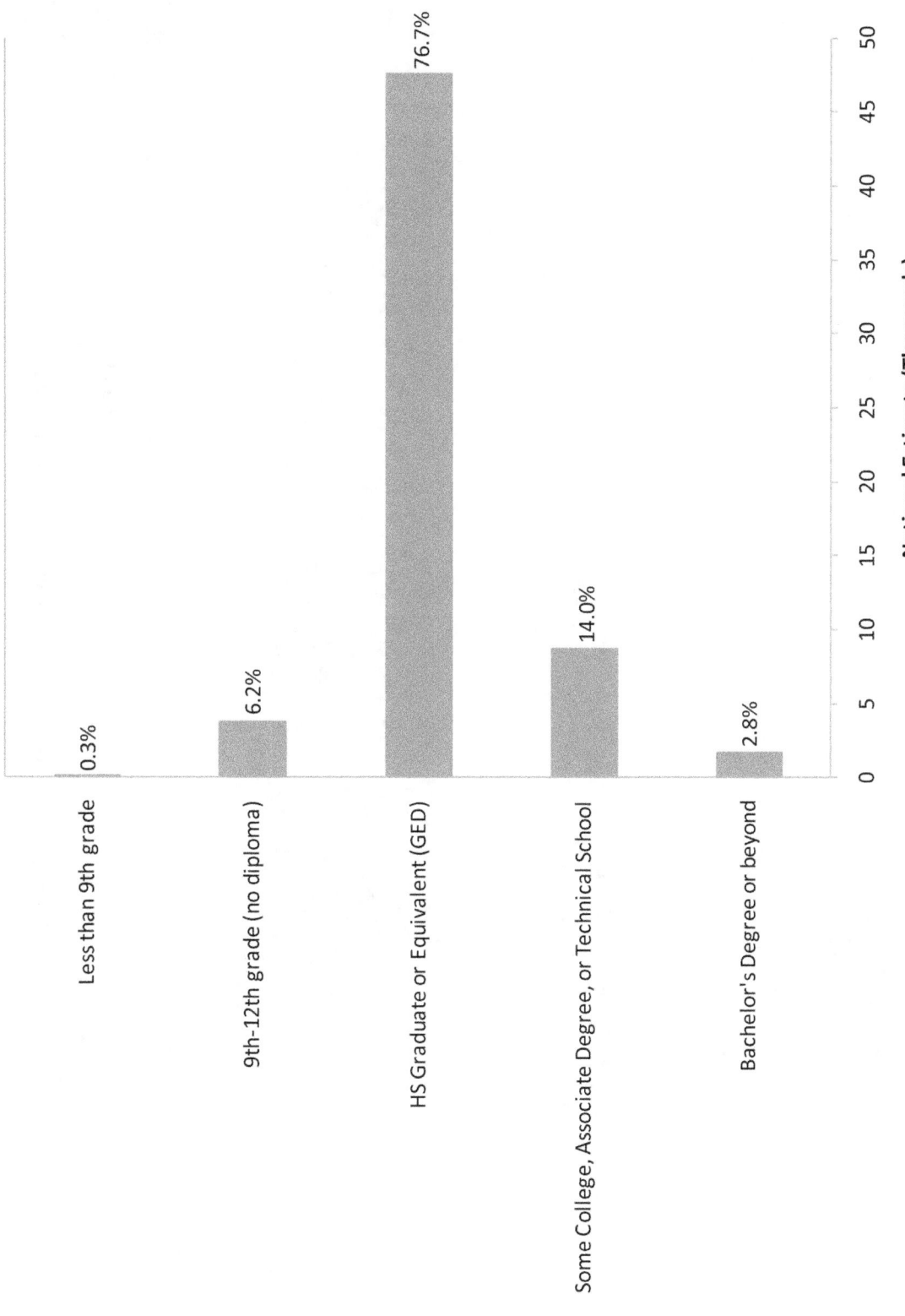

Figure 14. Education Level of Employees at Coal Mines.

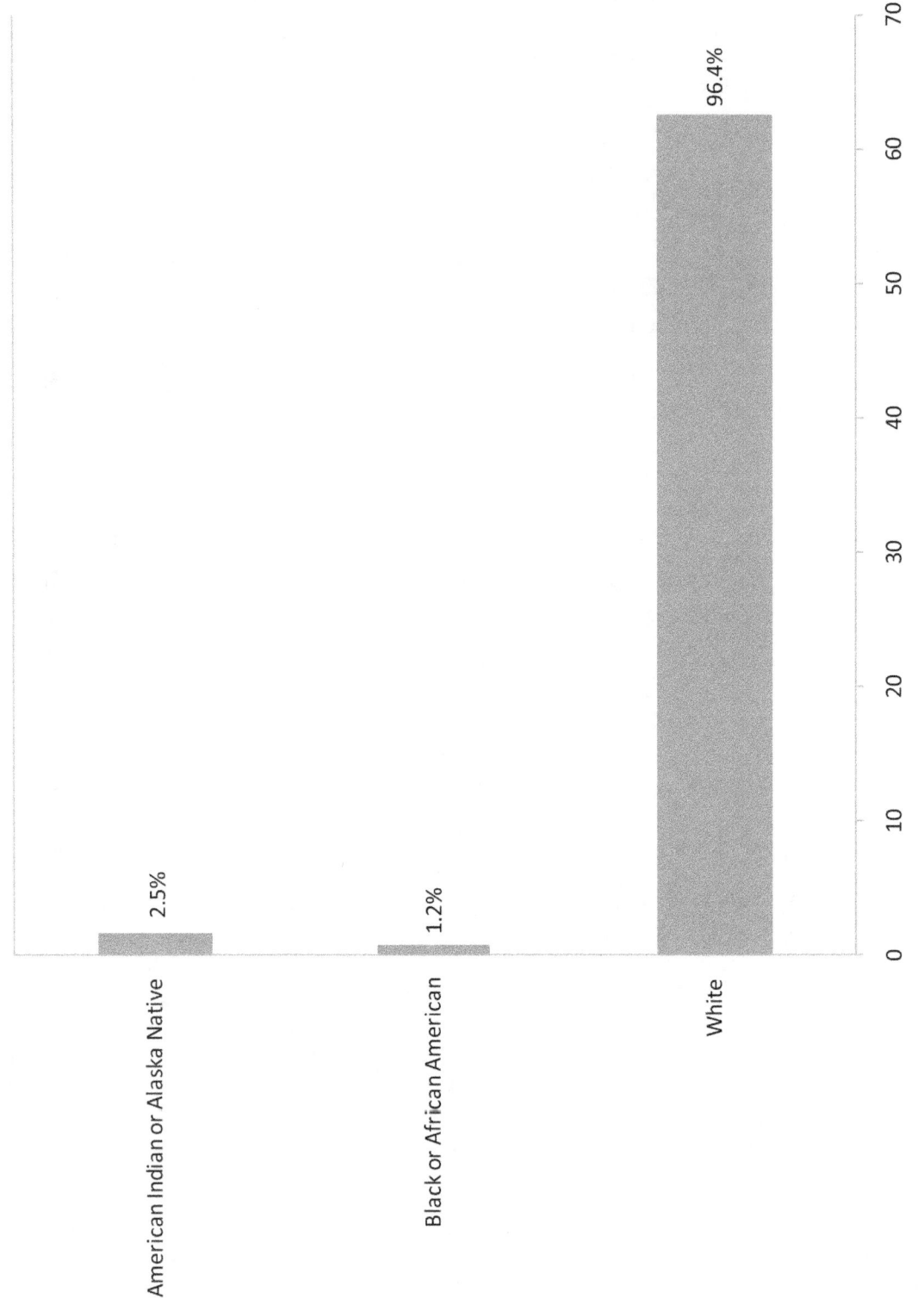

Figure 15. Race of Employees at Coal Mines.

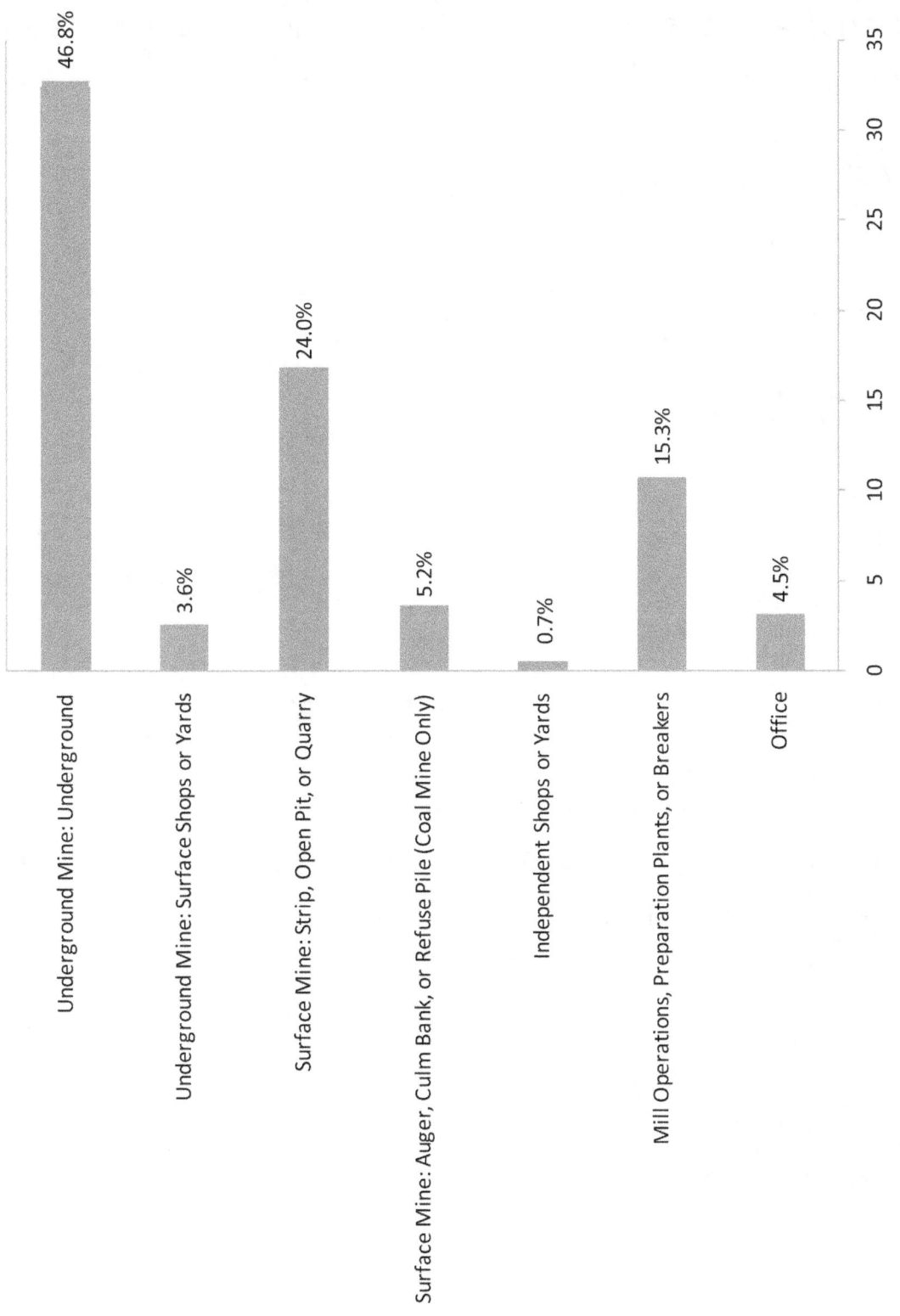

Figure 16. Primary Work Location of Employees at Coal Mines.

Table 25. Estimated Number of Administration/Professional Employees at Coal Mines

Occupation by Category	Survey Count	National Estimate	95% LCL	95% UCL
ADMINISTRATION/PROFESSIONAL	**523**	**16,048**	**12,552**	**19,544**
<u>Office Staff</u>	<u>69</u>	<u>2,120</u>	<u>1,395</u>	<u>2,846</u>
Administrative Staff	43	1,323	730	1,916
Administration				
Administrative Assistant				
Clerk				
Coal Distribution Coordinator				
Human Resources				
Mine Clerk				
Office Staff				
Secretary				
Technical Coordinator				
Business	16	479	254	704
Accounting				
Purchasing				
Sales				
Shipping				
Security	2	DSU	DSU	DSU
Supplies	8	254	49	459
Supply Clerk				
Warehouse Technician				
Warehouseman				
<u>Professional</u>	<u>59</u>	<u>2,214</u>	<u>683</u>	<u>3,746</u>
Engineer	10	303	42	564
Engineer (Electrical/Mining/Ventilation)				
Engineer, not otherwise specified				
Non-engineer	7	176	7	344
Environmental				
Professional, not otherwise specified				
Surveyor				
Technician	42	1,736	178	3,293
Coal Sampler				
Electronic Technician				
Engineering Technician				
Fuel Operator/Technician				
Lab Technician				
Plant Technician				
Technician				

Table 25. Estimated Number of Administration/Professional Employees at Coal Mines (continued)

Occupation by Category	Survey Count	National Estimate	95% LCL	95% UCL
<u>Safety</u>	<u>12</u>	<u>464</u>	<u>143</u>	<u>785</u>
Safety				
Safety Director				
Safety Supervisor				
<u>Supervisory</u>	<u>383</u>	<u>11,249</u>	<u>8,753</u>	<u>13,745</u>
Executive	*3*	*DSU*	*DSU*	*DSU*
CEO				
Mine Owner				
Foreman	*208*	*5,346*	*4,296*	*6,397*
Assistant Superintendent				
Belt Foreman (underground)				
Electrical Foreman (underground)				
Foreman				
Foreman/Manager				
Labor Foreman				
Lead Man				
Maintenance Foreman				
Outby Foreman				
Pit Foreman				
Preparation Plant Foreman				
Section Foreman				
Section Foreman/Boss				
Shift Foreman				
Superintendent				
Track Foreman				
Underground Foreman				
Manager	*76*	*3,187*	*1,456*	*4,918*
Assistant Manager				
Assistant Mine Foreman/Assistant Mine Manager				
Engineer/Operations Manager				
Maintenance Manager				
Management				
Manager				
Mine Foreman/Mine Manager				
Office Manager				
Plant Manager				
Plant Superintendent				

Table 25. Estimated Number of Administration/Professional Employees at Coal Mines (continued)

Occupation by Category	Survey Count	National Estimate	95% LCL	95% UCL
Supervisor	***96***	***2,657***	***1,677***	***3,636***
Assistant Supervisor				
Auger Crew Supervisor				
Belt Coordinator				
Electrical Supervisor				
Engineering Supervisor				
Maintenance Supervisor				
Mine Operator				
Mine Supervisor				
Pit Operator				
Pit Supervisor				
Plant Operator				
Plant Supervisor				
Prep Plant Operator				
Production Supervisor				
Supervisor				
Warehouse Supervisor				

Abbreviation: DSU, data suppressed.

Table 26. Estimated Number of Maintenance Employees at Coal Mines

Occupation by Category	Survey Count	National Estimate	95% LCL	95% UCL
MAINTENANCE	**370**	**12,000**	**8,929**	**15,071**
<u>Specialty</u>	<u>118</u>	<u>3,719</u>	<u>2,569</u>	<u>4,869</u>
Electrician	98	3,137	2,073	4,202
Electrician				
Electrician Trainee				
Maintenance Electrician				
Master Electrician				
Trainer Electrician				
Welder	20	582	205	959
Welder				
Welder (nonshop)				
Welder/Fabricator				
Welder/Mechanic				
<u>Support</u>	<u>252</u>	<u>8,281</u>	<u>5,764</u>	<u>10,798</u>
Maintenance	51	1,550	763	2,337
Continuous Miner Maintenance				
Dragline Oiler				
Greaser/Oiler				
Maintenance				
Maintenance Technician				
Mechanic Clerk				
Pipefitter				
Underground Belt Maintenance				
Underground Maintenance				
Mechanic	184	6,334	4,071	8,597
Belt Mechanic				
Diesel Mechanic				
Mechanic				
Mechanic/Electrician				
Mechanic Helper				
Mechanic Trainee				
Mobile Equipment Mechanic				
Plant Mechanic				
Prep Plant Mechanic				
Shop Mechanic				
Underground Belt Mechanic				
Repairman	17	397	76	717
Automotive Repairman				
Repairman				
Underground Belt Repairman				
Underground Repairman				

Table 27. Number of Miscellaneous Employees at Coal Mines

Occupation by Category	Survey Count
MISCELLANEOUS	15
Trainee	14
Unknown	1

Table 28. Estimated Number of Production Employees at Coal Mines

Occupation by Category	Survey Count	National Estimate	95% LCL	95% UCL
PRODUCTION	**1,016**	**29,562**	**23,638**	**35,485**
Equipment Operator	626	18,710	14,430	22,990
Dragline Operator	*11*	*369*	*21*	*718*
Equipment Operator	276	7,391	5,196	9,587
Backhoe Operator				
Bulldozer Operator				
Crane Operator				
End Dump Driver/Operator				
Equipment Operator				
Front End Loader				
Heavy Equipment Operator				
Highlift Operator				
Large Shovel/Backhoe/Load Operator				
Machine Operator				
Mobile Bridge Operator				
Mobile Equipment Operator				
Road Grader Operator				
Rock Duster				
Rotary Bucket Excavator Operator				
Rotary Dump Operator				
Scraper Operator				
Stationary Equipment Operator				
Tractor Operator/Motorman				
Hoist	2	DSU	DSU	DSU
Hoistman				
Material Mover	227	6,423	4,140	8,707
Dump Truck Driver				
Haul Truck Operator				
Hauler Operator				
Off Road Truck Driver				
Refuse Truck Driver/Backfill Truck Driver				
Rock Truck Driver				
Rubber Tire Operator				
Scoop Car Operator				
Shuttle Car Operator				

Table 28. Estimated Number of Production Employees at Coal Mines (continued)

Occupation by Category	Survey Count	National Estimate	95% LCL	95% UCL
Truck Driver				
Underground Coal Hauler				
Underground Haulage Operator				
Water Truck Operator				
Mining Machines	*97*	*3,906*	*2,484*	*5,328*
Continuous Miner Helper				
Continuous Miner Operator				
Face Operator				
Jacksetter				
Longwall Operator				
Shearer Operator				
Operator/Driver	*12*	*354*	*61*	*647*
Motorman				
Motorman/Locomotive Operator				
Transportation				
Shovel Operator	*1*	*DSU*	*DSU*	*DSU*
Extraction Labor	**53**	**1,609**	**256**	**2,963**
Coal Miner				
Mine Spec				
Miner Support				
Production Miner				
Material Preparation	**5**	**116**	**0**	**302**
Crusher	*4*	*DSU*	*DSU*	*DSU*
Crusher Attendant				
Cutter	*1*	*DSU*	*DSU*	*DSU*
Cutting Machine Operator				
Process	**13**	**334**	**85**	**584**
Conveyor Operator	*5*	*96*	*0*	*230*
Belt Vulcanizer				
Separation	*4*	*DSU*	*DSU*	*DSU*
Flotation Plant Operator				
Froth Cell Operator				
Wash Process	*3*	*DSU*	*DSU*	*DSU*
Washer Operator				
Wet Process	*1*	*DSU*	*DSU*	*DSU*
Wet Plant Attendant				
Support	**319**	**8,791**	**6,547**	**11,036**
Drill Operator	*25*	*616*	*245*	*986*
Auger Operator				
Coal Drill Operator				
Highwall Drill Operator				

Table 28. Estimated Number of Production Employees at Coal Mines (continued)

Occupation by Category	Survey Count	National Estimate	95% LCL	95% UCL
Explosives	15	638	130	1,145
Blaster				
Driller/Blaster				
Explosives/Powder Man				
Shooter				
Shot Firer				
Other	126	3,349	1,777	4,920
Control Man				
Dispatcher				
Operator, not otherwise specified				
Underground Operator				
Underground Plant Operator				
Quality Control	1	DSU	DSU	DSU
Roof Bolter	152	4,169	2,927	5,411

Abbreviation: DSU, data suppressed.

Table 29. Estimated Number of Service and Utility Employees at Coal Mines

Occupation by Category	Survey Count	National Estimate	95% LCL	95% UCL
SERVICE and UTILITY	410	11,791	9,398	14,184
<u>General Labor</u>	<u>163</u>	<u>4,863</u>	<u>3,101</u>	<u>6,625</u>
Cleaner	3	DSU	DSU	DSU
Cleanup Man				
Janitor				
Steamer				
Construction	2	DSU	DSU	DSU
Laborer	138	4,229	2,472	5,985
Inside Laborer				
Laborer				
Outby Laborer				
Outside Laborer				
Production Support				
Production Worker				
Shopman				
Surface Support				
Track Man				
Material Handling	4	DSU	DSU	DSU
Rolling Stock Crew				
Yard Man				
Tradesman	5	102	0	248
Apprentice/Journeyman				

Table 29. Estimated Number of Service and Utility Employees at Coal Mines (continued)

Occupation by Category	Survey Count	National Estimate	95% LCL	95% UCL
Weighman	11	239	73	404
Weighman				
Weighmaster				
<u>**Support Labor**</u>	<u>247</u>	<u>**6,928**</u>	<u>**4,770**</u>	<u>**9,085**</u>
Barge Operations	12	180	0	383
Barge Attendant/Boat Operator				
Boat Pilot				
Deck Hand				
Dock Hand				
Conveyor Operator	45	1,254	511	1,996
Belt Cleaner/Conveyor Man				
Belt Man/Conveyor Man				
Examiner	34	1,006	408	1,605
Fire Boss				
Mine Examiner				
Underground Belt Examiner				
Loading	74	1,514	988	2,040
Bin Attendant				
Loader Operator				
Loadout Operator				
Tipple Operator				
Underground Loader				
Pumper	9	400	0	902
Supplies	8	151	34	267
Parts Runner				
Supply Man				
Utility	61	2,298	1,043	3,552
Outside Utility				
Utility Belts				
Utility Bolter				
Utility Man				
Ventilation	4	DSU	DSU	DSU
Brattice Man				
Ventilation Man				

Abbreviation: DSU, data suppressed.

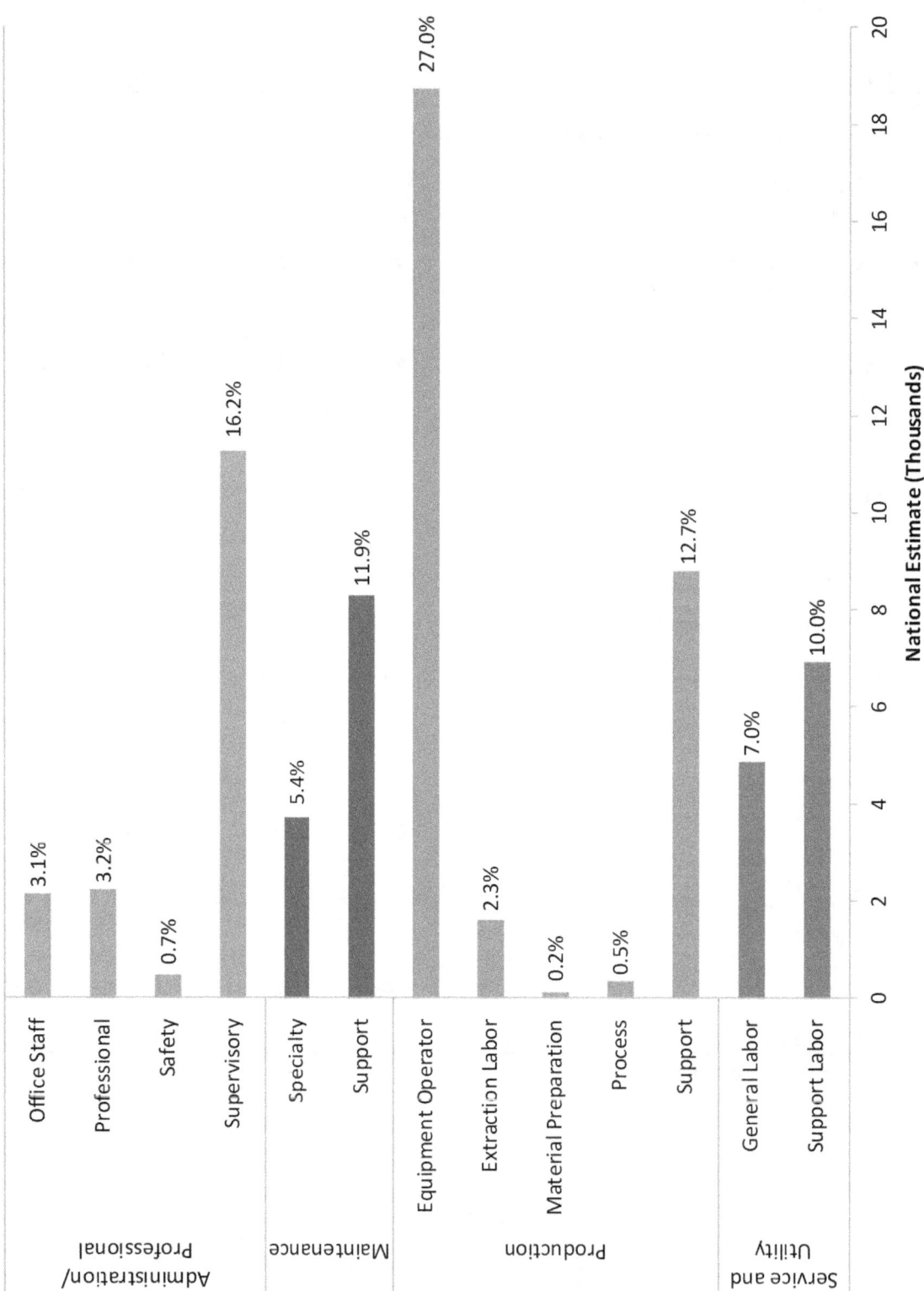

Figure 17. Occupational Categories of Employees at Coal Mines.

Employee Statistics for Metal Mines

Summary of Employee Statistics for Metal Mines

The demographic and occupational characteristics of employees in the U.S. metal mining industry are presented in Tables 30 and 31 and Figures 18–20. The weighted estimate for gender indicates that the workforce is composed predominately of male employees (86.2 percent). The majority of metal mine workers are White (91.4 percent), with another 4.7 percent of the workers having a racial category of Black or African American. Twenty-five percent of these employees are Hispanic or Latino. An estimated 53.3 percent are high school graduates and 43.6 percent have a level of education beyond high school. A review of the weighted estimates indicates that the average metal miner is 41.5 years of age and has worked in mining for 10.7 years, 8.7 years at the current mine, and 4.7 years in his/her job title. The national estimate for the average number of hours worked per week is 42.7. The primary work location for an estimated 42.7 percent of metal mine employees is a "Surface Mine: Strip, Open Pit or Quarry." An additional 23.9 percent of these employees work in "Mill Operations, Preparation Plants, or Breakers," while another 15.8 percent are employed in the "Underground Mine: Underground" work location.

Tables 32, 33, 35, 36, and Figure 21 present the national estimates of the number of workers by four major occupational categories. (No estimates were calculated for Table 34: "Miscellaneous.") An estimated 10,652 (27.5 percent) metal mine workers are employed in the "Administration/Professional" category; 7,238 (18.7 percent) in the "Maintenance" category; 17,581 (45.3 percent) in the "Production" category; and 3,339 (8.6 percent) in the "Service and Utility" category.

Table 30. Demographic Characteristics of Employees at Metal Mines

Demographic Characteristic	Survey Count	National Estimate	95% LCL	95% UCL	National Percent	95% LCL	95% UCL
Gender:							
Male	871	33,562	15,620	51,504	86.2	81.9	90.4
Female	93	5,383	1,152	9,615	13.8	9.6	18.1
Age (years)	958	41.5	39.3	43.8			
Highest level of education:							
Less than 9th grade	7	63	0	153	0.2	0.0	0.4
9th–12th grade (no diploma)	32	1,030	276	1,784	2.9	0.9	4.9
HS Graduate or Equivalent (GED)	496	18,934	9,552	28,317	53.3	44.0	62.6
Some College, Associate Degree, or Technical School	242	12,377	4,629	20,125	34.9	27.2	42.5
Bachelor's Degree or beyond	87	3,104	1,515	4,692	8.7	6.1	11.3
Ethnicity:							
Hispanic or Latino	137	9,483	1,132	17,834	24.6	14.4	34.9
Non-Hispanic or Non-Latino	783	29,008	14,213	43,803	75.4	65.1	85.6
Race:							
American Indian or Alaska Native	17	1,073	0	2,156	3.3	1.3	5.4
Asian	0	NA	NA	NA	NA	NA	NA
Black or African American	35	1,492	0	3,059	4.7	0.6	8.7
Native Hawaiian or Other Pacific Islander	3	DSU	DSU	DSU	DSU	DSU	DSU
White	818	29,276	16,297	42,255	91.4	86.8	96.0

Abbreviations: DSU, data suppressed; NA, not applicable.

Table 31. Occupational Characteristics of Employees at Metal Mines

Occupational Characteristic	Survey Count	National Estimate	95% LCL	95% UCL	National Percent	95% LCL	95% UCL
Hours worked (per week)	922	42.7	41.4	44.0			
Experience:							
Experience in this Job Title (years)	916	4.7	2.9	6.5			
Experience at this Mine (years)	928	8.7	7.3	10.0			
Total Mining Experience (years)	871	10.7	9.4	12.0			
Primary Work Location:							
Underground Mine: Underground	172	6,152	876	11,428	15.8	2.0	29.5
Underground Mine: Surface Shops or Yards	53	1,252	327	2,177	3.2	0.6	5.8
Surface Mine: Strip, Open Pit, or Quarry	204	16,624	0	34,516	42.7	20.9	64.4
Surface Mine: Dredge	1	DSU	DSU	DSU	DSU	DSU	DSU
Surface Mine: Other Surface Mining (Metal/Nonmetal Only)	127	1,876	405	3,348	4.8	0.5	9.1
Mill Operations, Preparation Plants, or Breakers	301	9,307	4,644	13,970	23.9	9.1	38.7
Office	106	3,751	782	6,720	9.6	6.9	12.4

Abbreviation: DSU, data suppressed.

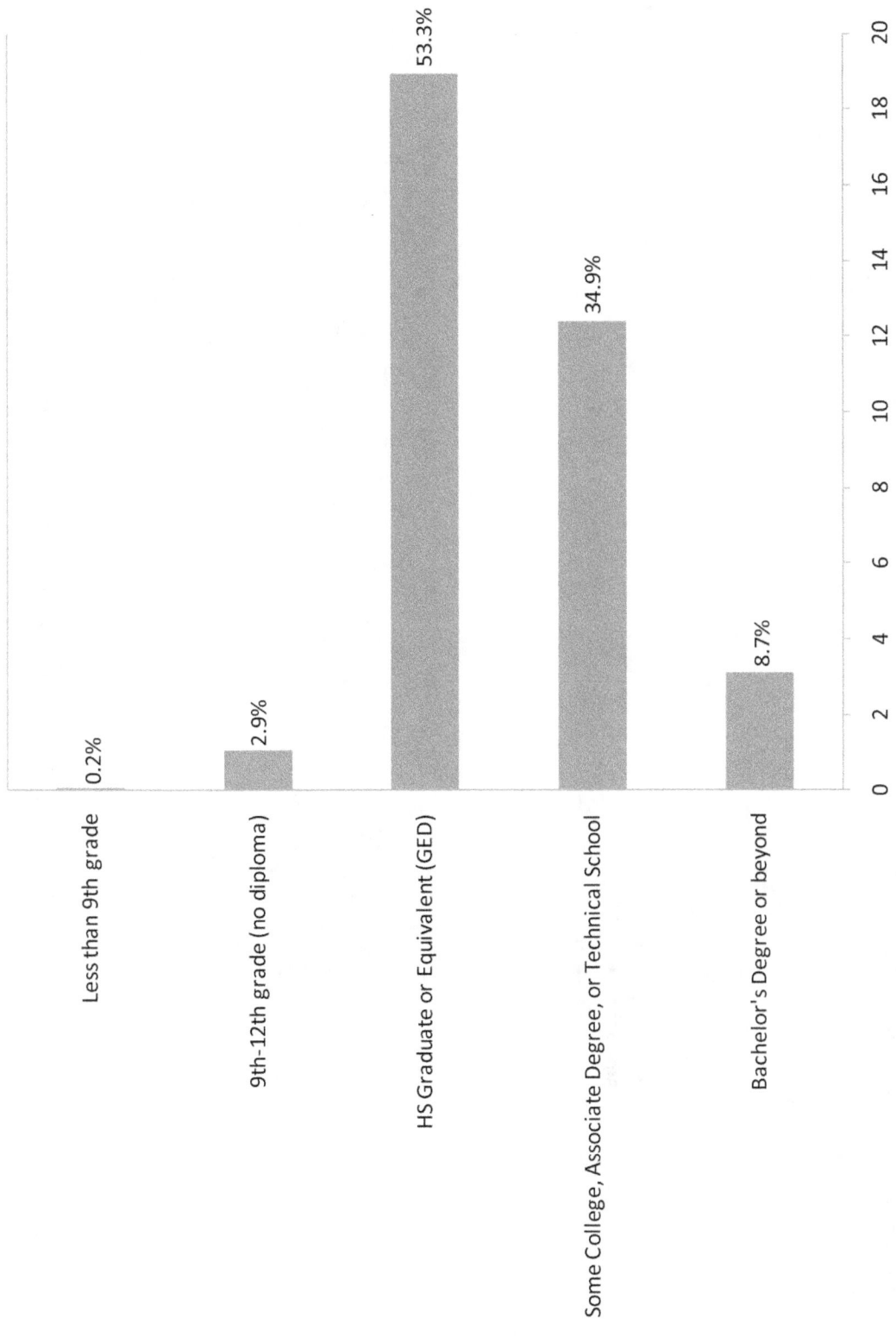

Figure 18. Education Level of Employees at Metal Mines.

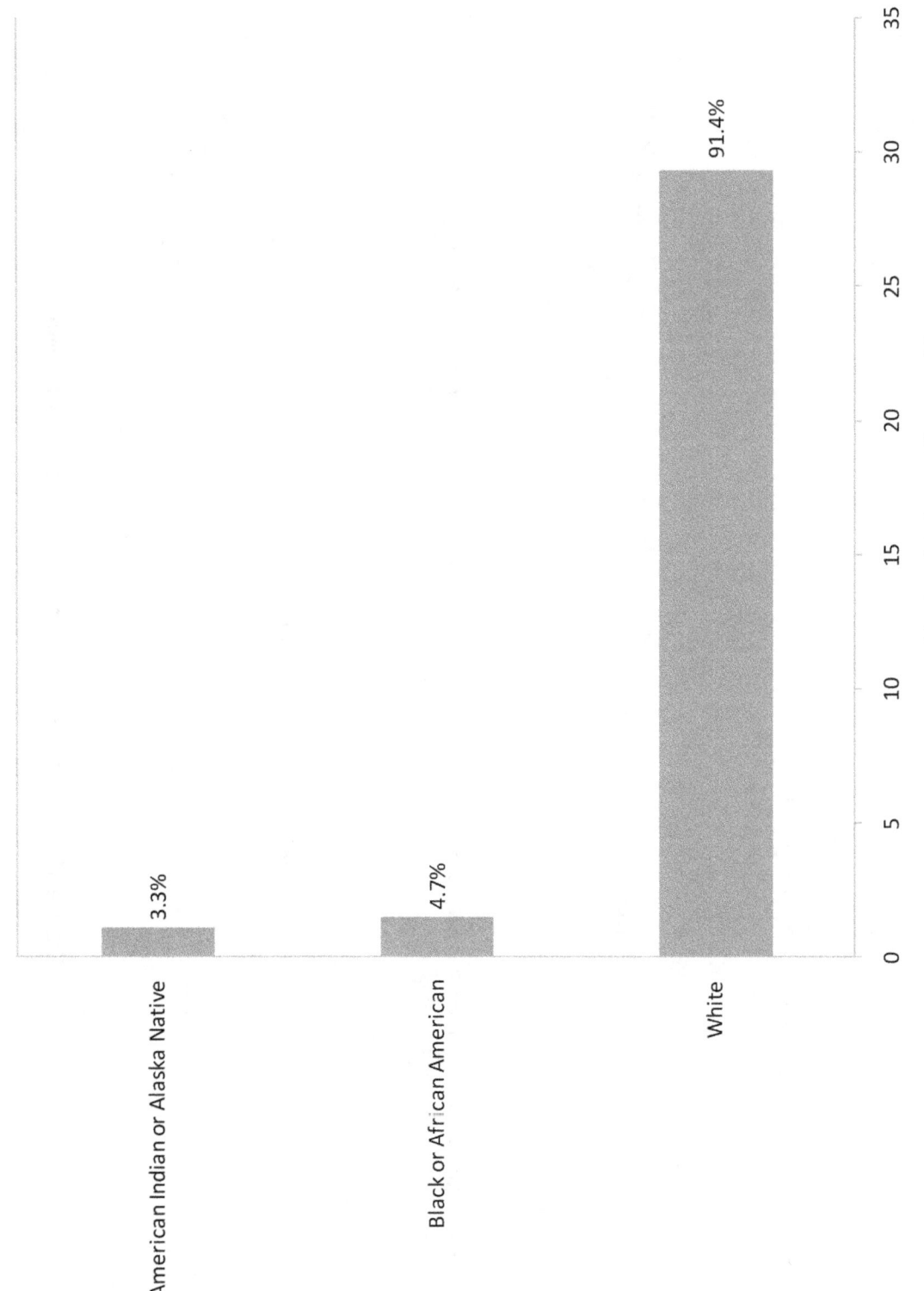

Figure 19. Race of Employees at Metal Mines.

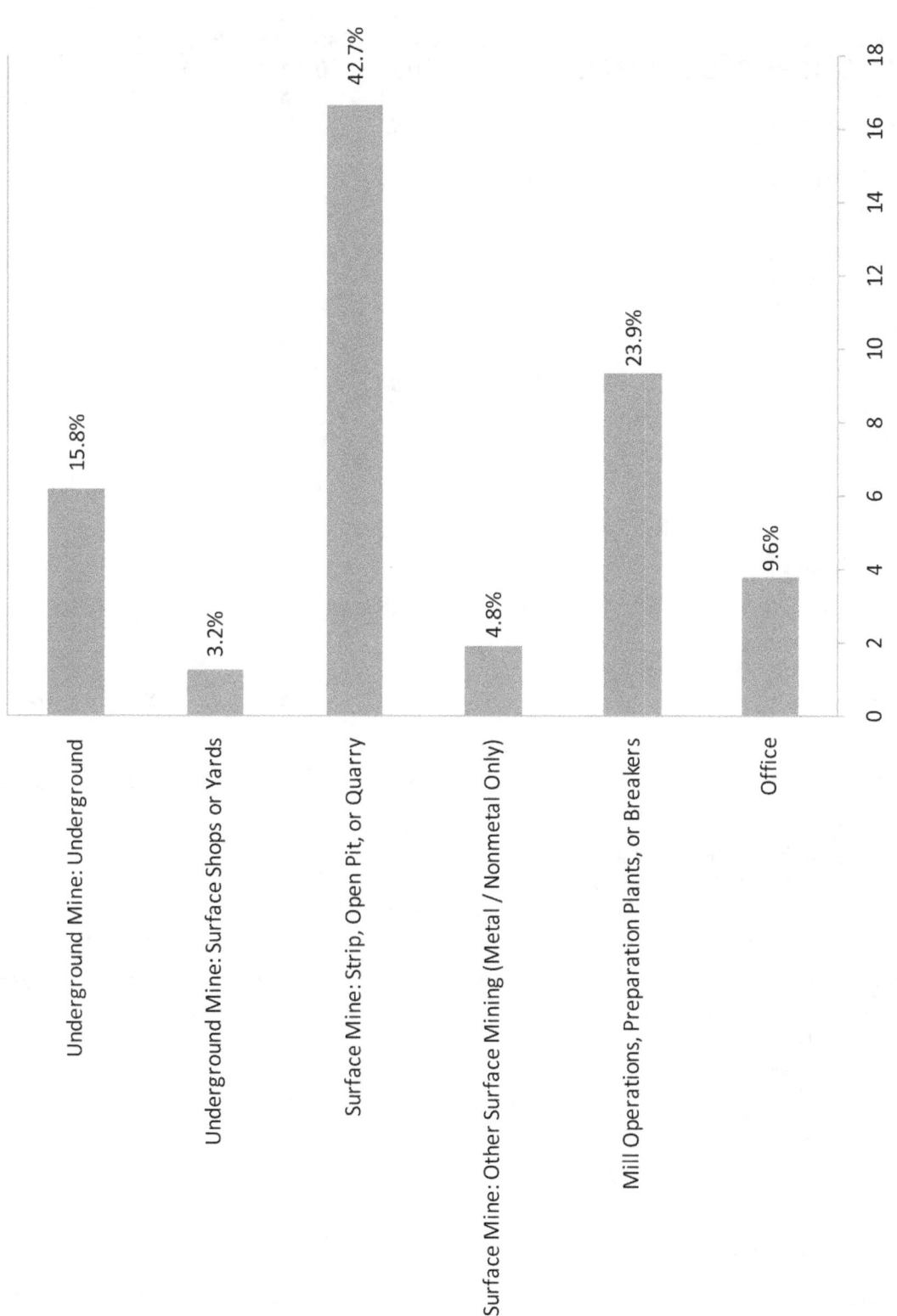

Figure 20. Primary Work Location of Employees at Metal Mines.

Table 32. Estimated Number of Administration/Professional Employees at Metal Mines

Occupation by Category	Survey Count	National Estimate	95% LCL	95% UCL
ADMINISTRATION/PROFESSIONAL	**308**	**10,652**	**5,754**	**15,550**
<u>Office Staff</u>	<u>49</u>	<u>1,889</u>	<u>725</u>	<u>3,053</u>
Administrative Staff	*24*	*811*	*241*	*1,381*
Administration				
Administrative Assistant				
Clerk				
Human Resources				
Office Staff				
Receptionist				
Secretary				
Systems Analyst				
Business	*16*	*804*	*0*	*1,782*
Accounting				
Bookkeeper				
Buyer				
Cost Coordinator				
Payroll				
Purchasing				
Shipping				
Security	*5*	*96*	*0*	*222*
Guard				
Supplies	*4*	*DSU*	*DSU*	*DSU*
Warehouse				
<u>Professional</u>	<u>85</u>	<u>3,368</u>	<u>1,260</u>	<u>5,475</u>
Engineer	*12*	*496*	*0*	*1,131*
Engineer (Electrical/Mining/Ventilation)				
Engineer, not otherwise specified				
Environmental Engineer				
Plant Engineer				
Non-engineer	*42*	*2,027*	*342*	*3,712*
Environmental Specialist				
Geologist				
Metallurgist				
Operations				
Operations Specialist				
Planner				
Professional, not otherwise specified				
Surveyor/Transit Man				
Utility Engineer				
Technician	*31*	*845*	*6*	*1,684*
Electrical Technician				
Laboratory Technician/Refiner				

Table 32. Estimated Number of Administration/Professional Employees at Metal Mines (continued)

Occupation by Category	Survey Count	National Estimate	95% LCL	95% UCL
Mechanic Technician				
Mill Technician				
Mine Technician				
Process Control Operator/Technician				
Sampler/Lab Technician				
Technician				
Utility Technician				
Safety	<u>11</u>	<u>303</u>	<u>87</u>	<u>519</u>
Safety				
Safety Manager				
Safety Supervisor				
Supervisory	<u>163</u>	<u>5,092</u>	<u>2,545</u>	<u>7,640</u>
Executive	*13*	*120*	*8*	*232*
CEO				
General Manager				
President				
Vice President				
Foreman	69	2,235	941	3,530
Assistant Superintendent				
Foreman				
Foreman/Shift Boss				
Lead Man				
Maintenance Foreman				
Mill Foreman				
Mine Foreman				
Plant Foreman				
Production Foreman				
Shift Foreman				
Superintendent				
Manager	26	410	85	735
Area Manager				
Concentrator Manager				
Engineering Manager				
Environmental Manager				
Human Resources Manager				
Manager				
Mill Manager				
Mine Manager				
Office Manager				
Plant Manager				
Process Manager				
Production Manager				
Project Manager				
Storeroom Manager				

Table 32. Estimated Number of Administration/Professional Employees at Metal Mines (continued)

Occupation by Category	Survey Count	National Estimate	95% LCL	95% UCL
Supervisor	*55*	*2,327*	*1,020*	*3,634*
Concentrator Supervisor				
Crusher Supervisor				
Gold House Supervisor				
Leaching Supervisor				
Maintenance Supervisor				
Mechanic Supervisor				
Mine Operations				
Mine Operator				
Mine Supervisor				
Plant Operator				
Process Supervisor				
Shift Supervisor				
Supervisor				
Tailings Supervisor				
Transportation Supervisor				
Warehouse Supervisor				

Abbreviation: DSU, data suppressed.

Table 33. Estimated Number of Maintenance Employees at Metal Mines

Occupation by Category	Survey Count	National Estimate	95% LCL	95% UCL
MAINTENANCE	179	7,238	3,058	11,418
<u>Specialty</u>	<u>28</u>	<u>1,585</u>	<u>350</u>	<u>2,819</u>
Electrician	22	1,483	264	2,702
Diagnostic Electrician				
Electrician/Wireman				
Welder	6	102	0	252
Maintenance Welder				
Welder				
<u>Support</u>	<u>151</u>	<u>5,653</u>	<u>2,515</u>	<u>8,791</u>
Maintenance	55	1,670	677	2,664
Crusher Maintenance				
Greaser/Oiler				
Maintenance				
Maintenance Planner				
Maintenance Technician				
Mill Maintenance				
Millwright				
Skilled Maintenance				
Mechanic	84	2,088	1,325	2,851
Automotive Mechanic				
Diagnostic Mechanic				
Diesel Mechanic				
Equipment Mechanic				
Heavy Equipment Mechanic				
Maintenance Mechanic				
Mechanic				
Mechanic/Welder				
Mechanic Helper				
Mobile Equipment Mechanic				
Mobile Maintenance Mechanic				
Plant Mechanic				
Repairman	12	1,895	0	4,796
Automotive Repairman				
Crusher Repairman				
Electronic/Electrical Repairman				
Heavy Duty Repairman				
Instrument Repairman				
Plant Repairman				
Repairman				
Tailings Repairman				

Table 34. Number of Miscellaneous Employees at Metal Mines

Occupation by Category	Survey Count
MISCELLANEOUS	3
Trainee	1
Unknown	2

Table 35. Estimated Number of Production Employees at Metal Mines

Occupation by Category	Survey Count	National Estimate	95% LCL	95% UCL
PRODUCTION	373	17,581	5,896	29,266
Equipment Operator	113	7,185	0	14,993
Equipment Operator	53	2,280	167	4,394
Bulldozer Operator				
Crane Operator				
Dredge Operator				
Equipment Operator				
Grader Operator				
Heavy Equipment Operator				
Mucking Machine Operator				
Raise Borer Operator				
Hoist	6	93	4	182
Hoist Operator				
Hoistman				
Skip Tender/Cager/Station Attendant				
Material Mover	42	3,569	0	7,682
Haul Truck Operator/Driver				
Truck Driver				
Mining Machines	4	DSU	DSU	DSU
Head Operator				
Shovel Operator	8	1,186	0	3,088
Extraction	60	2,192	119	4,265
Material Preparation	35	1,315	534	2,096
Additives	3	DSU	DSU	DSU
Additive Press Operator				
Thickener Operator				
Crusher	17	650	0	1,340
Crusher Helper				
Crusher Operator/Pan Feeder Operator				
Mill Crusher Operator				

Table 35. Estimated Number of Production Employees at Metal Mines (continued)

Occupation by Category	Survey Count	National Estimate	95% LCL	95% UCL
Mill	15	594	162	1,025
Mill Operator (ball/pebble/rod)				
Mill Production Worker				
Process	<u>54</u>	<u>3,088</u>	<u>78</u>	<u>6,098</u>
Conveyor Operator	3	DSU	DSU	DSU
Belt Vulcanizer				
Other	17	366	0	753
Process Operator				
Separation	34	2,360	0	5,284
Digestion Operator				
Filter Evaporation Operator				
Flotation/Concentrator Operator				
Grinder Operator				
Leach Utility				
Leaching Operations Worker				
Screen Plant Operator				
Tailings Operator				
Support	<u>111</u>	<u>3,801</u>	<u>503</u>	<u>7,099</u>
Drill Operator	17	237	57	416
Drill Operator				
Rotary Electric/Hydraulic Drill Operator				
Electronics	1	DSU	DSU	DSU
Power Systems				
Explosives	9	145	10	279
Blaster				
Driller/Blaster				
Other	75	3,212	0	6,461
Control Room				
Controller				
Dispatcher				
Operator, not otherwise specified				
Port Operator				
Production Operator				
Top Operator				
Quality Control	4	DSU	DSU	DSU
Quality Control/Quality Assurance				
Roof Bolter	5	42	0	104

Abbreviation: DSU, data suppressed.

Table 36. Estimated Number of Service and Utility Employees at Metal Mines

Occupation by Category	Survey Count	National Estimate	95% LCL	95% UCL
SERVICE and UTILITY	103	3,339	1,155	5,523
<u>General Labor</u>	<u>61</u>	<u>2,474</u>	<u>367</u>	<u>4,580</u>
Cleaners	1	DSU	DSU	DSU
Dry Attendant				
Construction	6	537	0	1,130
Cement Man/Concrete Worker				
Construction				
Shaft Miner/Shaft Repairer				
Laborer	30	160	14	307
Cook				
Laborer				
Production Worker				
Material Handling	5	156	0	390
Bagger/Bagging Operations Worker				
Material Handler				
Tradesman	18	1,590	0	3,617
Apprentice/Journeyman				
Boiler Operator				
Boilermaker				
Carpenter/Plumber/Painter				
Craftsman				
Trades Person				
Weighman	1	DSU	DSU	DSU
Weighmaster				
<u>Support Labor</u>	<u>42</u>	<u>865</u>	<u>253</u>	<u>1,478</u>
Conveyor Operator	4	DSU	DSU	DSU
Belt Cleaner/Conveyor Man				
Distribution	9	48	0	140
Packaging Operations Worker				
Loading	14	186	45	326
Chute Puller				
Load Haul Dump—Complete Cycle				
Loader Operator				
Supplies	5	39	0	78
Parts				
Supply Hauler				
Supply Man/Nipper				
Utility	10	380	0	890
Lampman				
Utility Man				

Abbreviation: DSU, data suppressed.

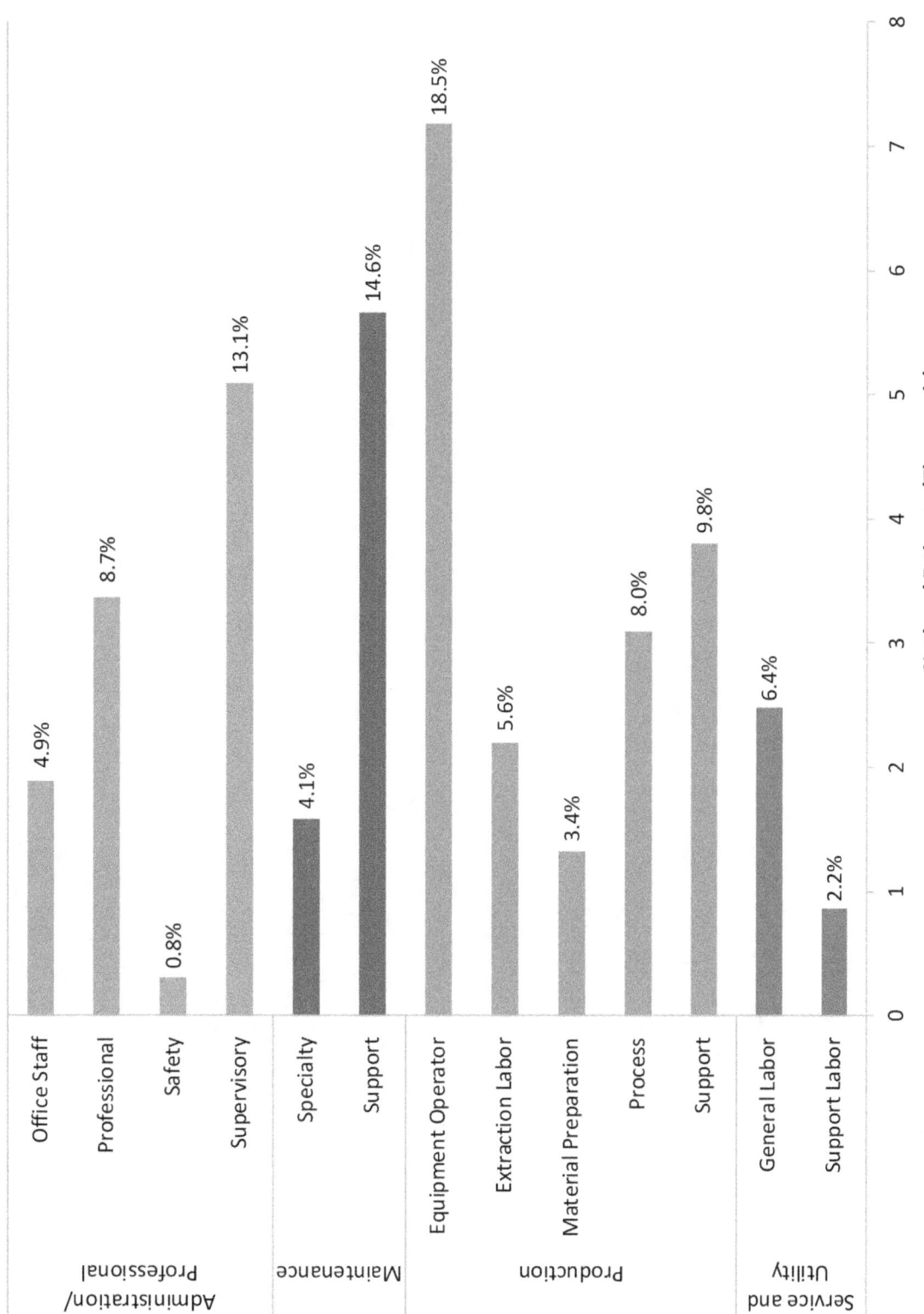

Figure 21. Occupational Categories of Employees at Metal Mines.

Employee Statistics for Nonmetal Mines

Summary of Employee Statistics for Nonmetal Mines

The demographic and occupational characteristics of employees in the U.S. nonmetal mining industry are presented in Tables 37 and 38 and Figures 22–24. The weighted survey estimate for gender indicates that the workforce is composed predominately of male employees (89.3 percent). The majority of nonmetal mine employees are White (85.6 percent) followed by Black or African American (13.6 percent). Eight percent of these employees have an ethnicity of Hispanic or Latino. An estimated 64.2 percent are high school graduates, with another 28.1 percent having a level of education beyond high school. A review of the weighted estimates indicates that the average nonmetal mine worker is 42.0 years of age and has worked in mining for 12.0 years, with 10.3 years at the current mine, and 6.7 years in his/her job title. The national estimate for the average number of hours worked per week is 42.4. The primary work location for an estimated 37.0 percent of nonmetal mine employees is "Mill Operations, Preparation Plants, or Breakers." An additional 24.4 percent of these employees work at a "Surface Mine: Other Surface Mining," while another 13.0 percent are employed in the "Surface Mine: Strip, Open Pit, or Quarry" work location.

Tables 39, 40, 42, 43, and Figure 25 present the national estimates of the number of nonmetal mine workers by four major occupational categories. (No estimates were calculated for Table 41: "Miscellaneous.") An estimated 7,066 (36.7 percent) are employed in the "Administration/Professional" category; 2,836 (14.7 percent) in the "Maintenance" category; 6,426 (33.3 percent) in the "Production" category; and 2,968 (15.4 percent) in the "Service and Utility" category.

Table 37. Demographic Characteristics of Employees at Nonmetal Mines

Demographic Characteristic	Survey Count	National Estimate	95% LCL	95% UCL	National Percent	95% LCL	95% UCL
Gender:							
Male	1,458	17,241	12,526	21,956	89.3	86.6	91.9
Female	136	2,074	1,174	2,973	10.7	8.1	13.4
Age (years)	1,505	42.0	40.2	43.8			
Highest level of education:							
Less than 9th grade	21	193	80	305	1.1	0.4	1.8
9th–12th grade (no diploma)	123	1,154	720	1,589	6.6	3.6	9.6
HS Graduate or Equivalent (GED)	888	11,242	6,837	15,647	64.2	58.1	70.3
Some College, Associate Degree, or Technical School	286	2,956	2,371	3,540	16.9	11.6	22.1
Bachelor's Degree or beyond	120	1,958	922	2,993	11.2	7.9	14.5
Ethnicity:							
Hispanic or Latino	158	1,368	854	1,883	8.3	5.7	10.9
Non-Hispanic or Non-Latino	1,384	15,171	12,851	17,491	91.7	89.1	94.3
Race:							
American Indian or Alaska Native	12	87	24	150	0.5	0.1	0.8
Asian	1	DSU	DSU	DSU	DSU	DSU	DSU
Black or African American	174	2,479	1,483	3,474	13.6	8.0	19.3
Native Hawaiian or Other Pacific Islander	3	DSU	DSU	DSU	DSU	DSU	DSU
White	1,262	15,567	10,412	20,721	85.6	79.8	91.4

Abbreviation: DSU, data suppressed.

Table 38. Occupational Characteristics of Employees at Nonmetal Mines

Occupational Characteristic	Survey Count	National Estimate	95% LCL	95% UCL	National Percent	95% LCL	95% UCL
Hours worked (per week)	1,489	42.4	41.6	43.3			
Experience:							
Experience in this Job Title (years)	1,570	6.7	5.3	8.1			
Experience at this Mine (years)	1,581	10.3	8.9	11.7			
Total Mining Experience (years)	1,507	12.0	10.9	13.2			
Primary Work Location:							
Underground Mine: Underground	175	1,971	1,050	2,892	10.3	5.5	15.0
Underground Mine: Surface Shops, Yards	31	422	94	751	2.2	0.5	3.9
Surface Mine: Strip, Open Pit, or Quarry	310	2,483	1,515	3,450	13.0	6.9	19.0
Surface Mine: Dredge	7	49	0	130	0.3	0.0	0.7
Surface Mine: Other Surface Mining (Metal/Nonmetal Only)	199	4,673	0	9,870	24.4	5.2	43.6
Independent Shops or Yards	16	159	5	313	0.8	0.0	1.7
Mill Operations, Preparation Plants, or Breakers	632	7,088	4,880	9,296	37.0	24.0	50.0
Office	213	2,324	1,502	3,146	12.1	7.2	17.0

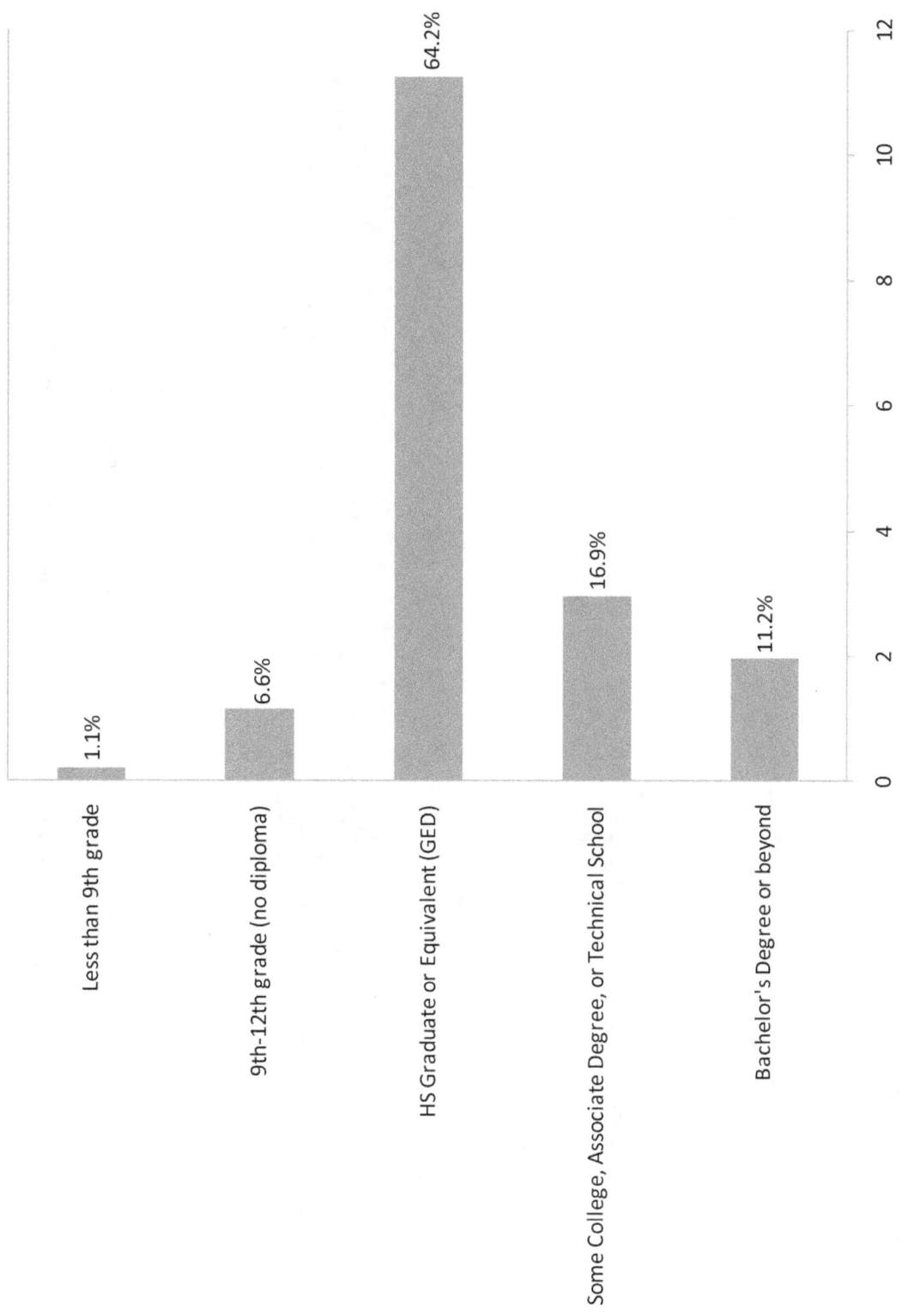

Figure 22. Education Level of Employees at Nonmetal Mines.

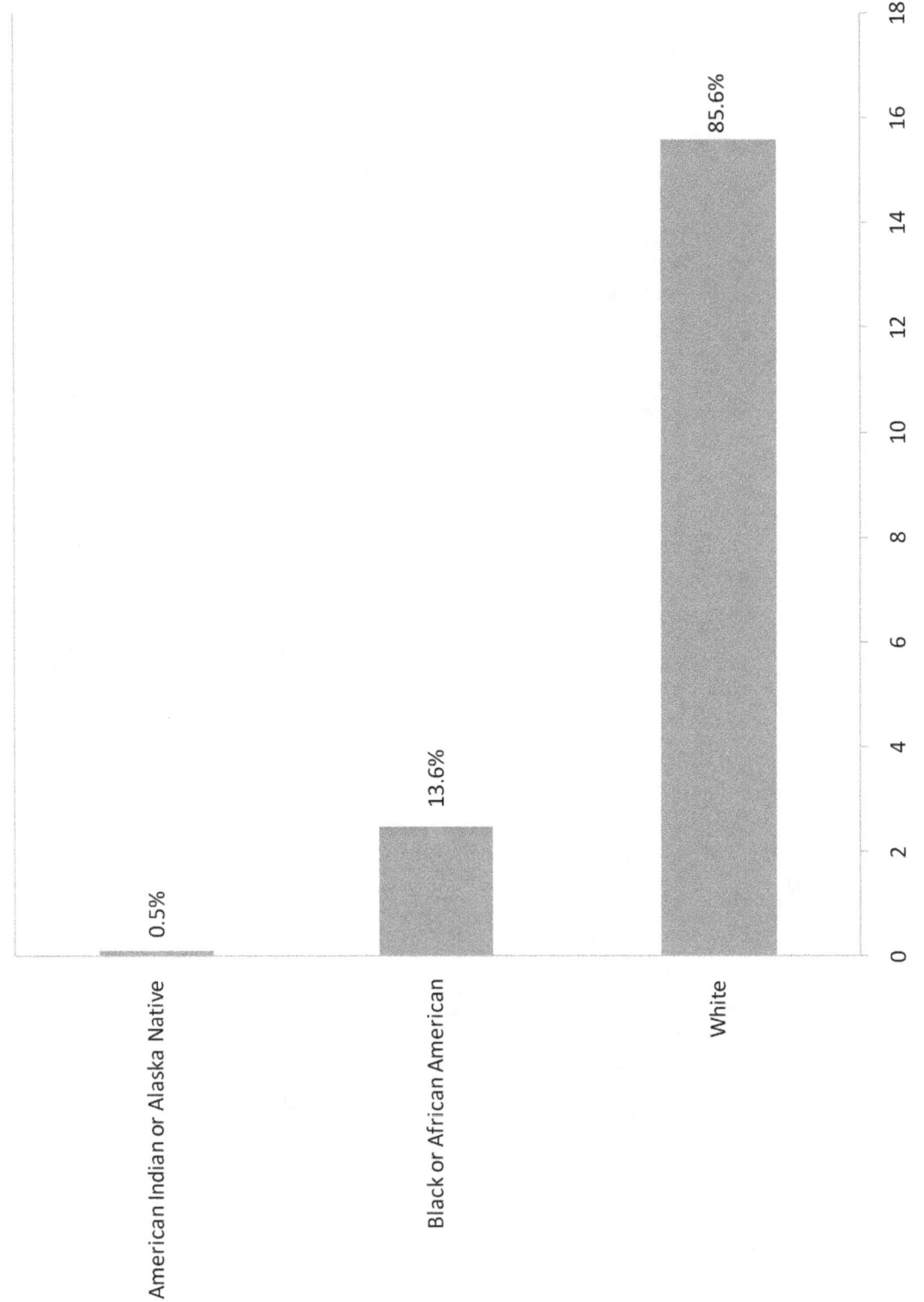

Figure 23. Race of Employees at Nonmetal Mines.

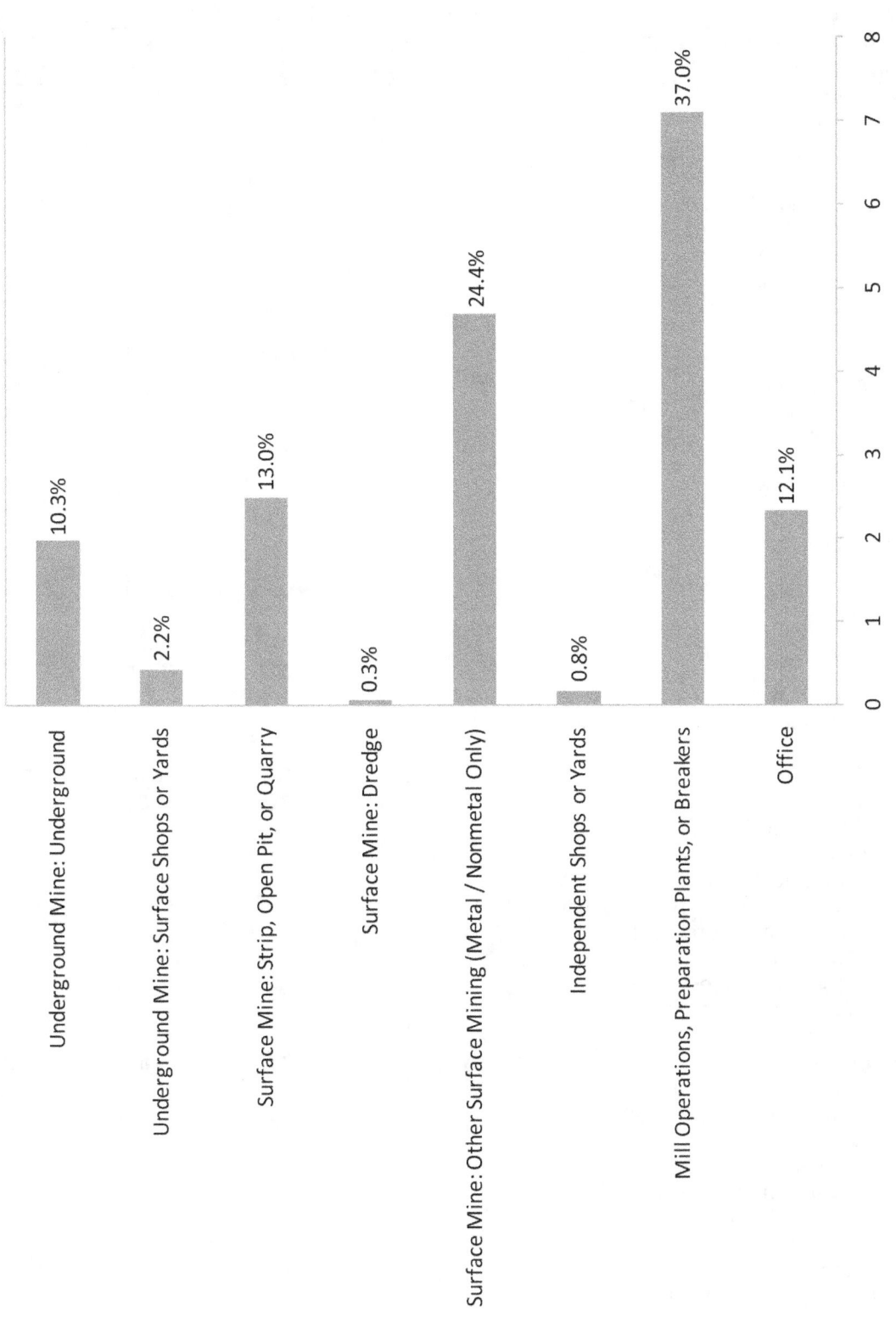

Figure 24. Primary Work Location of Employees at Nonmetal Mines.

Table 39. Estimated Number of Administration/Professional Employees at Nonmetal Mines

Occupation by Category	Survey Count	National Estimate	95% LCL	95% UCL
ADMINISTRATION/PROFESSIONAL	**499**	**7,066**	**3,597**	**10,536**
<u>Office Staff</u>	<u>108</u>	<u>1,504</u>	<u>901</u>	<u>2,107</u>
Administrative Staff	46	514	292	735
Administration				
Administrative Assistant				
Clerk				
Customer Service				
Human Resources				
Office Staff				
Receptionist				
Secretary				
Business	49	745	327	1,164
Accounting				
Bookkeeper				
Buyer				
Payroll				
Purchasing				
Sales				
Shipping				
Security	1	DSU	DSU	DSU
Guard				
Supplies	12	227	0	463
Supply Clerk				
Warehouse				
<u>Professional</u>	<u>65</u>	<u>987</u>	<u>618</u>	<u>1,355</u>
Engineer	18	297	84	509
Director of Engineering				
Engineer (Electrical/Mining/Ventilation)				
Engineer, not otherwise specified				
Environmental Engineer				
Plant Engineer				
Process Engineer				
Project Engineer				
Non-engineer	13	309	61	557
Control Person/Analyst				
Environmental Specialist				
Geologist				
Planner				
Production Scheduler				
Professional, not otherwise specified				
Reliability Engineer				
Surveyor/Transit Man				

Table 39. Estimated Number of Administration/Professional Employees at Nonmetal Mines (continued)

Occupation by Category	Survey Count	National Estimate	95% LCL	95% UCL
Technician	*34*	*381*	*182*	*581*
Electrical Technician				
Process Control Operator/Technician				
Sampler/Lab Technician				
Technician				
<u>**Safety**</u>	<u>7</u>	<u>147</u>	<u>0</u>	<u>314</u>
Inspector				
Safety				
<u>**Supervisory**</u>	<u>319</u>	<u>4,429</u>	<u>1,704</u>	<u>7,154</u>
Executive	*25*	*179*	*89*	*269*
General Manager				
Mine Owner				
President				
Vice President				
Foreman	*98*	*2,102*	*0*	*4,363*
Foreman				
Foreman/Shift Boss				
Lead Man				
Maintenance Foreman				
Maintenance Lead Man				
Mill Foreman				
Mine Foreman				
Pit Foreman				
Plant Foreman				
Production Foreman				
Shift Foreman				
Shop Foreman				
Superintendent				
Manager	*72*	*737*	*467*	*1,006*
Assistant Manager				
Customer Service Manager				
Environmental Manager				
Financial Manager				
Human Resources Manager				
Lab Manager				
Maintenance Manager				
Manager				
Mill Manager				
Mine Manager				
Office Manager				
Plant Manager				
Production Manager				
Project Manager				
Quality Control Manager				
Quarry Manager				
Raw Material Manager				
Sales Manager				

Table 39. Estimated Number of Administration/Professional Employees at Nonmetal Mines (continued)

Occupation by Category	Survey Count	National Estimate	95% LCL	95% UCL
Shipping Manager				
Supervisor	**124**	**1,411**	**828**	**1,995**
Assistant Mine Supervisor				
Bagging/Baghouse Supervisor				
Blasting Supervisor				
Clay Operator				
Electrical Supervisor				
Lab Supervisor				
Maintenance Supervisor				
Mine Operations				
Mine Operator				
Mine Supervisor				
Plant Operator				
Plant Supervisor				
Production Supervisor				
Quality Assurance Supervisor				
Quarry Supervisor				
Shift Supervisor				
Shipping Supervisor				
Supervisor				

Abbreviation: DSU, data suppressed.

Table 40. Estimated Number of Maintenance Employees at Nonmetal Mines

Occupation by Category	Survey Count	National Estimate	95% LCL	95% UCL
MAINTENANCE	**202**	**2,836**	**1,781**	**3,890**
<u>Specialty</u>	<u>30</u>	<u>437</u>	<u>168</u>	<u>706</u>
Electrician	25	401	136	667
Electrician/Wireman				
Maintenance Electrician				
Welder	5	35	0	81
Welder				
Welder/Mechanic				
<u>Support</u>	<u>172</u>	<u>2,399</u>	<u>1,531</u>	<u>3,267</u>
Maintenance	98	1,246	486	2,005
Electrical Maintenance				
Greaser/Oiler				
Maintenance				
Maintenance Clerk				
Maintenance Planner				
Maintenance Technician				
Mechanical Maintenance				
Millwright				
Plant Maintenance				
Road Maintenance				
Mechanic	70	1,099	554	1,644
Diesel Mechanic				
Heavy Equipment Mechanic				
Maintenance Mechanic				
Master Mechanic				
Mechanic				
Mechanic Helper				
Mobile Equipment Mechanic				
Mobile Maintenance Mechanic				
Mobile Mechanic				
Plant Mechanic				
Wrens Mechanic				
Repairman	4	DSU	DSU	DSU
Automotive Repairman				
Heavy Duty Repairman				
Maintenance Repairman				

Abbreviation: DSU, data suppressed.

Table 41. Number of Miscellaneous Employees at Nonmetal Mines

Occupation by Category	Survey Count
MISCELLANEOUS	**4**
Trainee	2
Unknown	2

Table 42. Estimated Number of Production Employees at Nonmetal Mines

Occupation by Category	Survey Count	National Estimate	95% LCL	95% UCL
PRODUCTION	**636**	**6,426**	**5,142**	**7,710**
Equipment Operator	221	1,892	1,308	2,477
Equipment Operator	*124*	*1,058*	*603*	*1,512*
Backhoe Operator				
Bulldozer Operator				
Crane Operator				
Dredge Operator				
Equipment Operator				
Forklift Operator				
Front End Loader Operator				
Grader Operator				
Gravity Mag Operator				
Heavy Equipment Operator				
Mobile Equipment Operator				
Rotary Bucket Excavator Operator				
Scraper Operator				
Stripping Operator				
Tractor Operator				
Hoist	*12*	*117*	*22*	*211*
Hoist Engineer				
Hoist Operator				
Hoistman				
Skip Tender/Cager/Station Attendant				
Material Mover	*75*	*548*	*284*	*812*
Haul Truck Operator/Driver				
Hauler/Haul Unit Operator				
Off Road Truck Driver				
Ore Truck Driver/Operator				
Pit Truck Driver				
Rock Truck Driver				
Scoop Loader				
Scoop Tram Operator				
Truck Driver				
Water Truck Operator				

Table 42. Estimated Number of Production Employees at Nonmetal Mines (continued)

Occupation by Category	Survey Count	National Estimate	95% LCL	95% UCL
Mining Machines	5	93	0	204
Continuous Miner Operator				
Head Operator				
Undercutter Operator				
Operator/Driver	3	DSU	DSU	DSU
Dump Operator				
Transportation				
Shovel Operator	2	DSU	DSU	DSU
<u>Extraction Labor</u>	<u>77</u>	<u>1,018</u>	<u>363</u>	<u>1,673</u>
Mine Production				
Mine Support				
Miner				
<u>Material Preparation</u>	<u>106</u>	<u>918</u>	<u>544</u>	<u>1,293</u>
Additives	10	121	0	286
Calcine Operator				
Crusher	30	180	83	276
Blunging Operator				
Crusher Helper				
Crusher Operator/Pan Feeder Operator				
Screenhouse Crusher				
Cutter	11	45	0	124
Cutting Machine Operator				
Sawyer				
Mill	55	572	271	873
Dry Mill Operator				
Mill Hand/Helper				
Mill Operator (ball/pebble/rod)				
Mill Production Worker				
Roller Mill Operator				
<u>Process</u>	<u>61</u>	<u>659</u>	<u>370</u>	<u>948</u>
Conveyor Operator	1	DSU	DSU	DSU
Belt Vulcanizer				
Dry Processing	20	199	56	342
Dry Plant/Process Operator				
Dryer Operator				
Kiln Operator				
Other	6	90	0	210
Fabricator				
Process Attendant				
Process Operator				

Table 42. Estimated Number of Production Employees at Nonmetal Mines (continued)

Occupation by Category	Survey Count	National Estimate	95% LCL	95% UCL
Separation	*28*	*337*	*132*	*541*
Centrifuge Utility				
Extruder Operator				
Filter Operator				
Flotation/Concentrator Operator				
Grinder Operator				
Leaching Operations Worker				
Mix Operator				
Pan Operator				
Screen Plant Labor				
Screen Plant Operator				
Slurry Operator				
Wet Process	*6*	*28*	*0*	*58*
Wet Plant Operator				
Support	**171**	**1,938**	**988**	**2,889**
Drill Operator	*11*	*72*	*21*	*123*
Electronics	*2*	*DSU*	*DSU*	*DSU*
Robot Operator				
Explosives	*8*	*175*	*0*	*406*
Blaster				
Driller/Blaster				
Explosives/Powder Man				
Other	*129*	*1,412*	*525*	*2,298*
Control Room				
Controller				
Operator, not otherwise specified				
Production Operator				
Rak Handler				
Quality Control	*15*	*186*	*51*	*320*
Quality Control/Quality Assurance				
Roof Bolter	*6*	*61*	*0*	*124*

Abbreviation: DSU, data suppressed.

Table 43. Estimated Number of Service and Utility Employees at Nonmetal Mines

Occupation by Category	Survey Count	National Estimate	95% LCL	95% UCL
SERVICE and UTILITY	**254**	**2,968**	**2,236**	**3,699**
<u>General Labor</u>	<u>150</u>	<u>1,776</u>	<u>1,217</u>	<u>2,334</u>
Cleaner	*2*	*DSU*	*DSU*	*DSU*
Janitor				
Tank Car Washer				
Construction	*16*	*198*	*18*	*378*
Construction				
Packer				
Shaft Miner/Shaft Repairer				
Laborer	*51*	*587*	*256*	*918*
Laborer				
Miller				
Plant Helper				
Plant Man				
Production Worker				
Quarry Worker				
Material Handling	*78*	*928*	*502*	*1,355*
Bagger/Bagging Operations Worker				
Crude Pile Operator				
Material Handler				
Palletizer				
Reclaim Operator				
Stacker				
Storage Operator				
Storeroom				
Yard Laborer				
Tradesman	*2*	*DSU*	*DSU*	*DSU*
Boiler Operator				
Weighman	*1*	*DSU*	*DSU*	*DSU*
Scale Clerk/Operator				
<u>Support Labor</u>	<u>104</u>	<u>1,192</u>	<u>742</u>	<u>1,641</u>
Barge Operations	*6*	*46*	*0*	*111*
Barge Attendant/Boat Operator				
Deck Hand				
Conveyor Operator	*1*	*DSU*	*DSU*	*DSU*
Belt Cleaner/Conveyor Man				
Distribution	*18*	*316*	*0*	*697*
Packaging Operations Worker				

Table 43. Estimated Number of Service and Utility Employees at Nonmetal Mines (continued)

Occupation by Category	Survey Count	National Estimate	95% LCL	95% UCL
Loading	*49*	*561*	*344*	*779*
Bulk Loader				
Load Haul Dump—Complete Cycle				
Loader Operator				
Loading				
Plant Loader Operator				
Production Loader				
Rail Loader Operator				
Shipping Loader				
Stock Loader/Piler				
Tipple Operator				
Utility	*30*	*252*	*79*	*425*
Operator Utility				
Plant Utility				
Quarry Utility				
Utility Lubricator				
Utility Man				

Abbreviation: DSU, data suppressed.

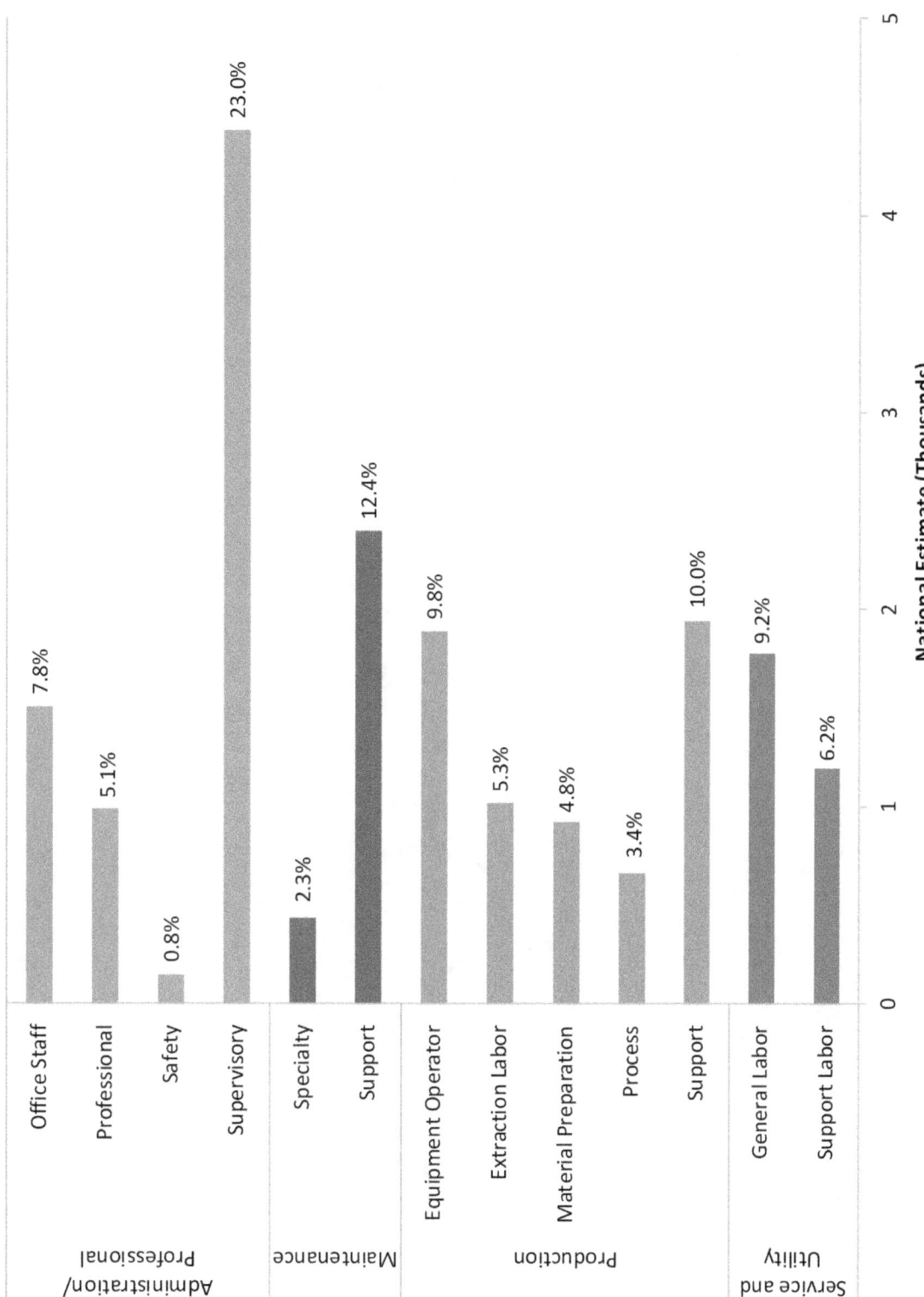

Figure 25. Occupational Categories of Employees at Nonmetal Mines.

Employee Statistics for Stone Mines

Summary of Employee Statistics for Stone Mines

The demographic and occupational characteristics of employees in the U.S. stone mining industry are presented in Tables 44 and 45 and Figures 26–28. The weighted estimate for gender indicates that the workforce is composed predominately of male employees (93.4 percent). The majority of stone mine workers are White (93.8 percent), with another 4.7 percent of the workers having a racial category of Black or African American. Fourteen percent of these employees are Hispanic or Latino. An estimated 62.0 percent are high school graduates and 20.9 percent have a level of education beyond high school. A review of the weighted estimates indicates that the average stone miner is 43.8 years of age and has worked in mining for 12.5 years, 10.3 years at the current mine, and 7.8 years in his/her job title. The national estimate for the average number of hours worked per week is 45.7. The primary work location for an estimated 36.3 percent of stone mine employees is a "Surface Mine: Strip, Open Pit or Quarry." An additional 33.5 percent of these employees work in "Mill Operations, Preparation Plants, or Breakers," while another 14.4 percent are employed in the "Surface Mine: Other Surface Mining" work location.

Tables 46, 47, 49, 50, and Figure 29 present the national estimates of the number of workers by four major occupational categories. (No estimates were calculated for Table 48: "Miscellaneous.") An estimated 19,435 (27.5 percent) stone mine workers are employed in the "Administration/Professional" category; 10,563 (14.9 percent) in the "Maintenance" category; 24,955 (35.3 percent) in the "Production" category; and 15,826 (22.3 percent) in the "Service and Utility" category.

Table 44. Demographic Characteristics of Employees at Stone Mines

Demographic Characteristic	Survey Count	National Estimate	95% LCL	95% UCL	National Percent	95% LCL	95% UCL
Gender:							
Male	2,545	65,950	60,931	70,970	93.4	92.3	94.4
Female	173	4,666	3,802	5,530	6.6	5.6	7.7
Age (years)	2,629	43.8	42.9	44.7			
Highest level of education:							
Less than 9th grade	111	3,094	1,630	4,558	4.7	2.5	6.9
9th–12th grade (no diploma)	320	8,075	6,195	9,956	12.4	9.6	15.1
HS Graduate or Equivalent (GED)	1,607	40,481	35,504	45,457	62.0	56.9	67.1
Some College, Associate Degree, or Technical School	353	10,020	7,927	12,112	15.3	12.6	18.1
Bachelor's Degree or beyond	129	3,647	2,686	4,607	5.6	4.2	6.9
Ethnicity:							
Hispanic or Latino	309	9,394	6,111	12,676	13.6	9.1	18.1
Non-Hispanic or Non-Latino	2,348	59,768	54,166	65,370	86.4	81.9	90.9
Race:							
American Indian or Alaska Native	25	815	306	1,323	1.3	0.5	2.1
Asian	4	DSU	DSU	DSU	DSU	DSU	DSU
Black or African American	104	3,040	1,551	4,529	4.7	2.5	7.0
Native Hawaiian or Other Pacific Islander	6	198	0	441	0.3	0.0	0.7
White	2,362	60,494	55,116	65,872	93.8	91.5	96.0

Abbreviation: DSU, data suppressed.

Table 45. Occupational Characteristics of Employees at Stone Mines

Occupational Characteristic	Survey Count	National Estimate	95% LCL	95% UCL	National Percent	95% LCL	95% UCL
Hours worked (per week)	2,601	45.7	44.2	47.2			
Experience:							
Experience in this Job Title (years)	2,596	7.8	7.2	8.3			
Experience at this Mine (years)	2,635	10.3	9.3	11.2			
Total Mining Experience (years)	2,643	12.5	11.7	13.3			
Primary Work Location:							
Underground Mine: Underground	217	1,710	1,305	2,115	2.4	1.8	3.0
Underground Mine: Surface Shops, Yards	121	732	482	983	1.0	0.7	1.4
Surface Mine: Strip, Open Pit, or Quarry	917	25,736	21,819	29,654	36.3	31.6	41.0
Surface Mine: Dredge	6	248	0	584	0.4	0.0	0.8
Surface Mine: Other Surface Mining (Metal/Nonmetal Only)	352	10,203	7,034	13,372	14.4	10.0	18.8
Independent Shops or Yards	22	530	0	1,077	0.7	0.0	1.5
Mill Operations, Preparation Plants, or Breakers	782	23,787	19,554	28,021	33.5	28.3	38.8
Office	301	7,957	6,438	9,475	11.2	9.3	13.1

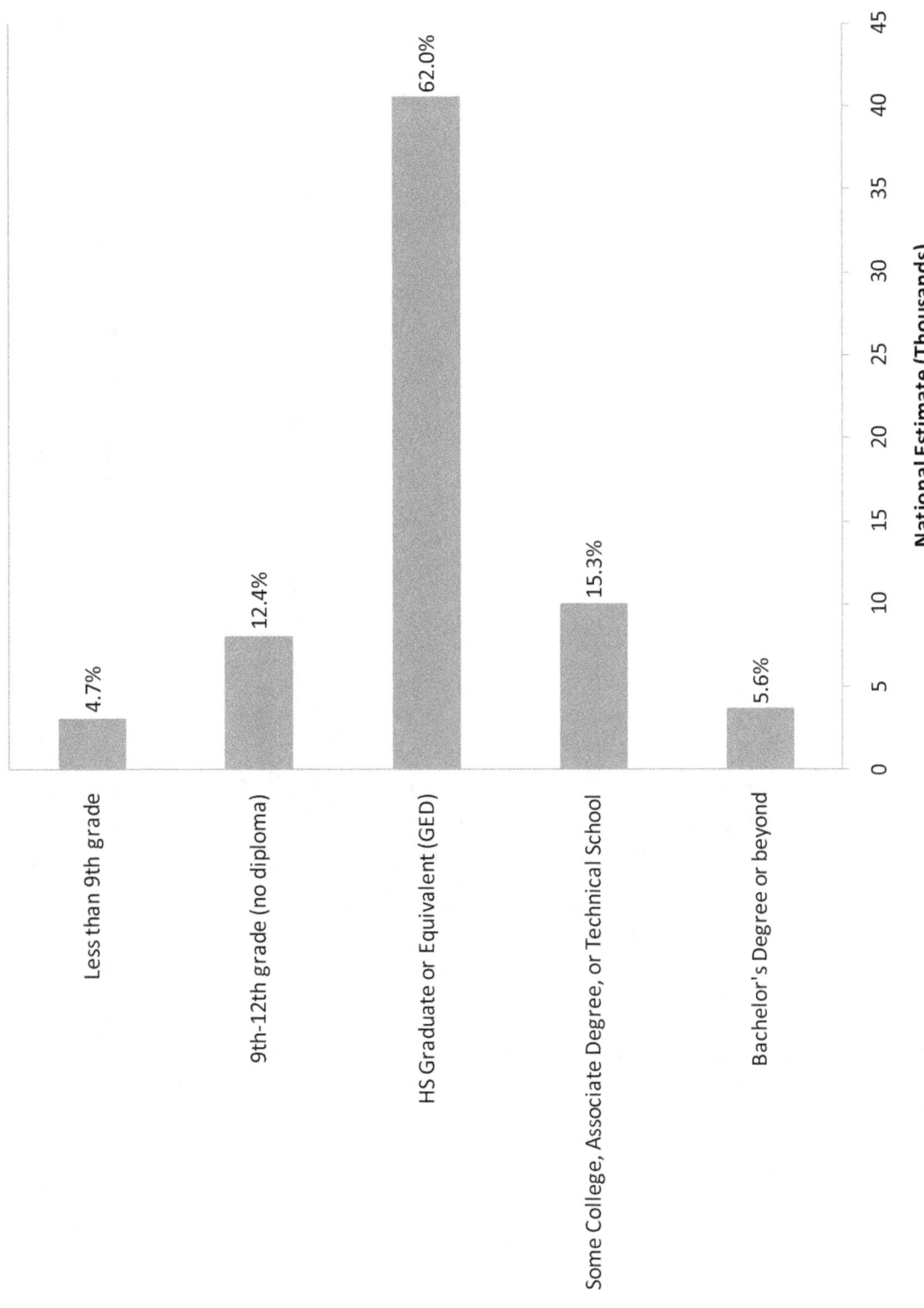

Figure 26. Education Level of Employees at Stone Mines.

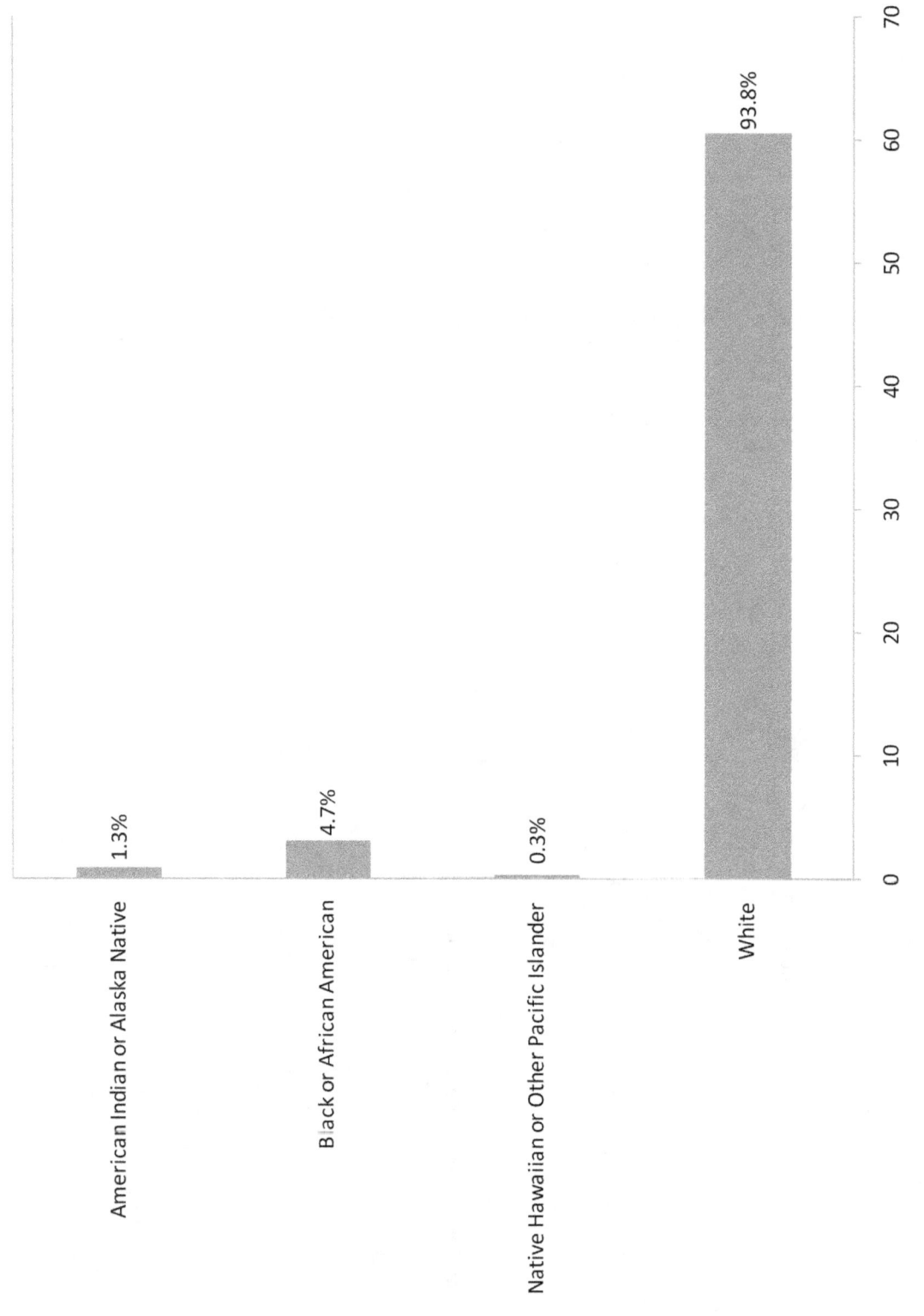

Figure 27. Race of Employees at Stone Mines.

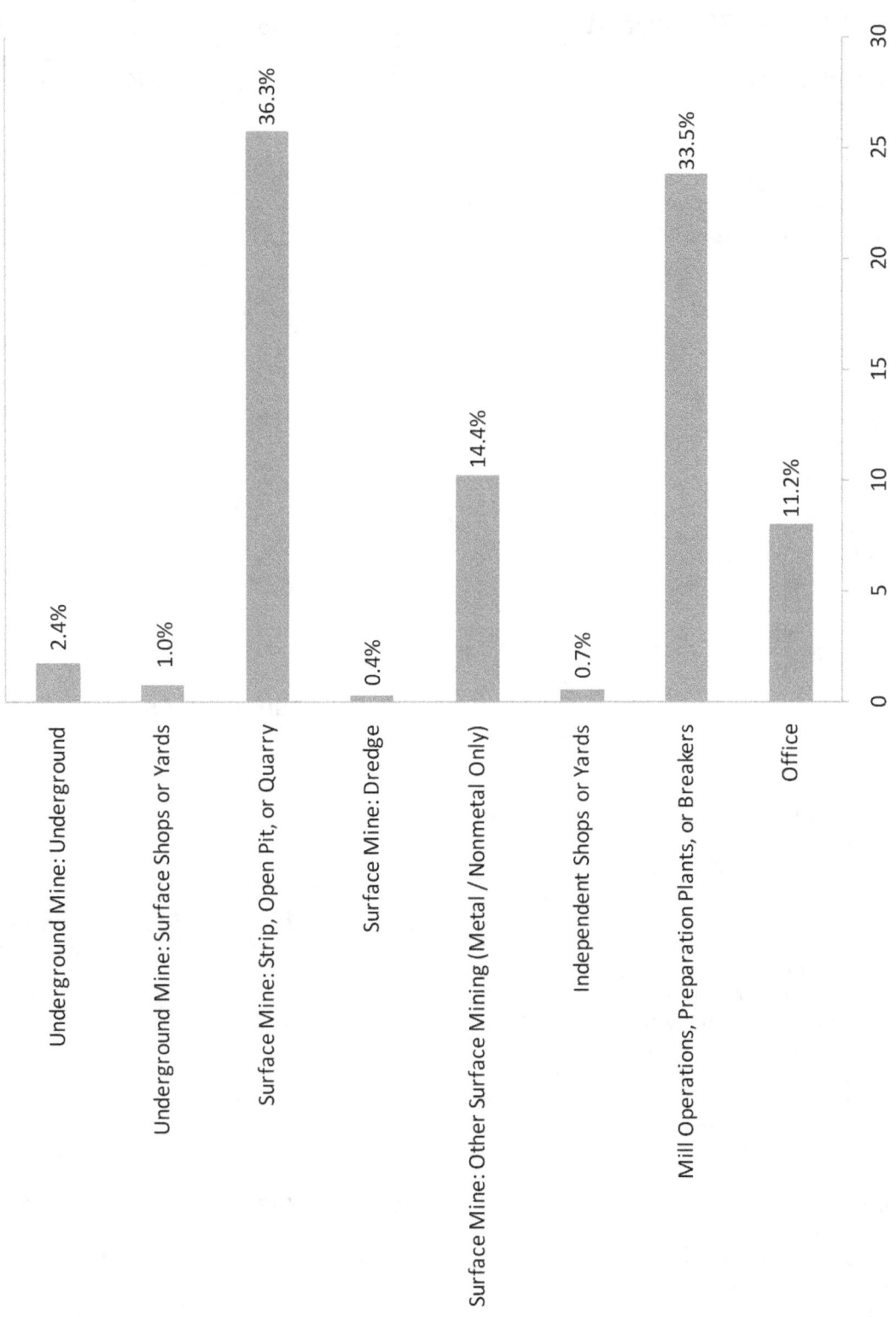

Figure 28. Primary Work Location of Employees at Stone Mines.

Table 46. Estimated Number of Administration/Professional Employees at Stone Mines

Occupation by Category	Survey Count	National Estimate	95% LCL	95% UCL
ADMINISTRATION/PROFESSIONAL	**725**	**19,435**	**16,978**	**21,891**
<u>Office Staff</u>	<u>117</u>	<u>3,155</u>	<u>2,320</u>	<u>3,990</u>
Administrative Staff	*69*	*1,748*	*1,250*	*2,245*
Administration				
Administrative Assistant				
Clerk				
Communications				
Customer Service				
Human Resources				
Information Technology				
Office Clerk				
Office Staff				
Plant Clerk				
Secretary				
Business	*38*	*1,131*	*609*	*1,653*
Accounting				
Bookkeeper				
Buyer				
Payroll				
Procurement				
Purchasing				
Sales				
Shipping				
Terminal Operator				
Security	*3*	*DSU*	*DSU*	*DSU*
Guard				
Supplies	*7*	*158*	*8*	*308*
Supply Clerk				
Warehouse				
Warehouse Technician				
<u>Professional</u>	<u>92</u>	<u>3,139</u>	<u>1,947</u>	<u>4,332</u>
Engineer	*14*	*566*	*234*	*897*
Engineer (Electrical/Mining/Ventilation)				
Engineer, not otherwise specified				
Environmental Engineer				
Plant Engineer				
Process Engineer				
Production Engineer				
Project Engineer				
Non-engineer	*16*	*596*	*166*	*1,025*
Chemist				
Control Person/Analyst				
Environmental Specialist				
Physical Tester				
Planner				

Table 46. Estimated Number of Administration/Professional Employees at Stone Mines (continued)

Occupation by Category	Survey Count	National Estimate	95% LCL	95% UCL
Professional, not otherwise specified				
Reliability Engineer				
Technician	62	1,978	990	2,966
Electrical Technician				
Materials Technician				
Process Control Operator/Technician				
Production Technician				
Quarry Technician				
Sampler/Lab Technician				
Technician				
Utility Technician				
Safety	<u>11</u>	<u>346</u>	<u>80</u>	<u>613</u>
Inspector				
Safety				
Safety Manager				
Supervisory	<u>505</u>	<u>12,794</u>	<u>11,322</u>	<u>14,265</u>
Executive	22	659	390	929
General Manager				
Mine Owner				
President				
Vice President				
Foreman	165	4,255	3,512	4,998
Assistant Superintendent				
Foreman				
Foreman/Shift Boss				
Lead Man				
Maintenance Foreman				
Maintenance Lead Man				
Mine Foreman				
Pit Foreman				
Plant Foreman				
Shop Foreman				
Superintendent				
Manager	105	2,406	1,880	2,933
Assistant Manager				
Distribution Manager				
Environmental Manager				
Equipment Maintenance Manager				
Equipment Manager				
Facility Manager				
Human Resources Manager				
Maintenance Manager				
Manager				
Mine Manager				
Office Manager				

Table 46. Estimated Number of Administration/Professional Employees at Stone Mines (continued)

Occupation by Category	Survey Count	National Estimate	95% LCL	95% UCL
Operations Manager				
Plant Manager				
Production Manager				
Project Manager				
Purchasing Manager				
Quality Control Manager				
Quarry Manager				
Regulatory Manager				
Sales Manager				
Scale Office Manager				
Technical Service Manager				
Supervisor	213	5,473	4,284	6,662
Blasting Supervisor				
Control Room Supervisor				
Crusher Supervisor				
Electrical Supervisor				
Equipment Supervisor				
Lab Supervisor				
Loader Supervisor				
Loadhouse Supervisor				
Maintenance Supervisor				
Mine Operator				
Mine Supervisor				
Mobile Equipment Supervisor				
Plant Operator				
Plant Supervisor				
Process Supervisor				
Production Supervisor				
Quality Assurance Supervisor				
Quarry Operator				
Quarry Supervisor				
Shift Supervisor				
Shipping Supervisor				
Supervisor				
Transportation Supervisor				
Warehouse Supervisor				
Wash Plant Supervisor				

Abbreviation: DSU, data suppressed.

Table 47. Estimated Number of Maintenance Employees at Stone Mines

Occupation by Category	Survey Count	National Estimate	95% LCL	95% UCL
MAINTENANCE	384	10,563	8,999	12,127
<u>Specialty</u>	<u>79</u>	<u>2,219</u>	<u>1,533</u>	<u>2,904</u>
Electrician	39	1,191	709	1,674
Electrician/Wireman				
Maintenance Electrician				
Welder	40	1,027	572	1,483
Maintenance Welder				
Repair/Welder				
Welder				
Welder/Mechanic				
<u>Support</u>	<u>305</u>	<u>8,344</u>	<u>6,908</u>	<u>9,780</u>
Maintenance	132	3,604	2,585	4,624
Electrical Maintenance				
Equipment Maintenance				
Fixed Maintenance				
Greaser/Oiler				
Liquid Fuel Handler				
Maintenance				
Maintenance Clerk				
Maintenance Coordinator				
Maintenance Planner				
Maintenance Technician				
Mechanical Maintenance				
Millwright				
Mobile Maintenance				
Plant Maintenance				
Mechanic	149	3,721	2,632	4,811
Heavy Equipment Mechanic				
Maintenance Mechanic				
Master Mechanic				
Mechanic				
Mechanic Helper				
Mobile Equipment Mechanic				
Mobile Maintenance Mechanic				
Mobile Mechanic				
Plant Mechanic				
Repairman	24	1,019	258	1,780
Automotive Repairman				
Electronic/Electrical Repairman				
Instrument Repairman				
Mechanical Repairman				
Repairman				

Table 48. Number of Miscellaneous Employees at Stone Mines

Occupation by Category	Survey Count
MISCELLANEOUS	7
Trainee	1
Unknown	6

Table 49. Estimated Number of Production Employees at Stone Mines

Occupation by Category	Survey Count	National Estimate	95% LCL	95% UCL
PRODUCTION	1,040	24,955	21,960	27,951
Equipment Operator	589	14,803	12,345	17,261
Dragline Operator	4	DSU	DSU	DSU
Equipment Operator	297	8,113	6,018	10,208
Bobcat Operator				
Bulldozer Operator				
Crane Operator				
Dredge Operator				
End Dump Driver				
Equipment Operator				
Forklift Operator				
Front End Loader Operator				
Grader Operator				
Heavy Equipment Operator				
Highlift Operator				
Machine Operator				
Mobile Equipment Operator				
Paver Operator				
Payloader Operator				
Rotary Bucket Excavator Operator				
Scaler (mechanical)				
Tower Operator				
Track Hoe				
Tractor Operator				
Material Mover	275	6,209	4,898	7,521
Dump Operator				
Haul Truck Operator/Driver				
Hauler/Haul Unit Operator				
Motorman				
Off Road Truck Driver				
Operator/Driver				
Pit Truck Driver				
Quarry Truck Driver				
Rock Truck Driver				
Stock Truck/Stock Pile Driver				
Transportation				
Truck Driver				
Water Truck Operator				

Table 49. Estimated Number of Production Employees at Stone Mines (continued)

Occupation by Category	Survey Count	National Estimate	95% LCL	95% UCL
Shovel Operator	3	DSU	DSU	DSU
Extraction Labor	**22**	**410**	**0**	**950**
Heading Prep				
Miner				
Material Preparation	**115**	**2,718**	**1,728**	**3,709**
Additives	*1*	*DSU*	*DSU*	*DSU*
Additives Utility				
Crusher	*45*	*978*	*586*	*1,369*
Breaker Operator				
Crusher Operator/Pan Feeder Operator				
Crusher Plant Operator				
Hammer Mill Operator				
Jaw Operator				
Rock Breaker Operator				
Cutter	*47*	*1,115*	*215*	*2,015*
Sawyer				
Splitter				
Stone Cutter				
Trimmer				
Mill	*22*	*547*	*167*	*927*
Limestone Prep Operator				
Mill Man				
Mill Operator (ball/pebble/rod)				
Milling Machine Operator				
Roller Operator				
Process	**34**	**1,177**	**588**	**1,767**
Dry Processing	*10*	*373*	*33*	*713*
Dryer Operator				
Kiln Operator				
Other	*9*	*369*	*50*	*689*
Fabricator				
Process Attendant				
Separation	*15*	*435*	*93*	*776*
Grinder Operator				
Mix Chemist				
Mix Operator				
Pelletizing Operations Worker				
Pug Operator/Mixer Tender				
Rotex Operator				
Screen Plant Operator				

Table 49. Estimated Number of Production Employees at Stone Mines (continued)

Occupation by Category	Survey Count	National Estimate	95% LCL	95% UCL
<u>Support</u>	<u>280</u>	<u>5,847</u>	<u>4,387</u>	<u>7,308</u>
Drill Operator	46	730	442	1,019
Drill Helper/Chuck Tender				
Drill Operator				
Electronics	1	DSU	DSU	DSU
Console Operator				
Explosives	48	561	288	834
Blaster				
Driller/Blaster				
Explosives/Powder Man				
Shooter				
Other	142	3,500	2,176	4,823
Control Room				
Controller				
Dispatcher				
Operator, not otherwise specified				
Panel Operator				
Production Operator				
Scaler (hand)				
Quality Control	38	965	638	1,292
Quality Control/Quality Assurance				
Roof Bolter	5	49	0	107
Roof Bolter				
Roof Control Operator				

Abbreviation: DSU, data suppressed.

Table 50. Estimated Number of Service and Utility Employees at Stone Mines

Occupation by Category	Survey Count	National Estimate	95% LCL	95% UCL
SERVICE and UTILITY	565	15,826	13,213	18,439
<u>General Labor</u>	<u>293</u>	<u>9,020</u>	<u>6,871</u>	<u>11,169</u>
Cleaner	3	DSU	DSU	DSU
Janitor				
Tower Cleaner				
Construction	11	271	0	547
Curb Cutter				
Ground Control/Timberman				
Packer				
Screed Person				
Laborer	190	6,107	4,029	8,184
Ground Man				
Laborer				
Miller				
Plant Man				
Quarry Worker				
Material Handling	36	1,215	518	1,911
Bagger/Bagging Operations Worker				
Material Handler				
Palletizer				
Silo Operator				
Stacker				
Storeroom				
Yard Laborer				
Tradesman	5	141	0	319
Apprentice/Journeyman				
Machinist				
Weighman	48	1,183	836	1,529
Scale Clerk/Operator				
Weighmaster				
<u>Support Labor</u>	<u>272</u>	<u>6,806</u>	<u>5,172</u>	<u>8,440</u>
Barge Operations	6	188	0	407
Barge Attendant/Boat Operator				
Deck Hand				
Dock Worker				
Conveyor Operator	2	DSU	DSU	DSU
Belt Cleaner/Conveyor Man				
Distribution	7	187	0	407
Packaging Operations Worker				
Packhouse				

Table 50. Estimated Number of Service and Utility Employees at Stone Mines (continued)

Occupation by Category	Survey Count	National Estimate	95% LCL	95% UCL
Loading	*202*	*4,840*	*3,396*	*6,284*
Bin Puller/Truck Loader				
Bulk Loader				
Load Man				
Loader Operator				
Loading				
Loadout Operator				
Pit Loader Operator				
Plant Loader Operator				
Production Loader				
Quarry Loader Operator				
Rail Loader Operator				
Stock Loader/Piler				
Yard Loader Operator				
Utility	55	1,561	912	2,210
Crusher Utility				
E.O. Utility				
Equipment Utility				
Mill Utility				
Pit Utility Person				
Plant Utility				
Production Utility				
Quarry Utility				
Utility Man				
Utility Scaler				

Abbreviation: DSU, data suppressed.

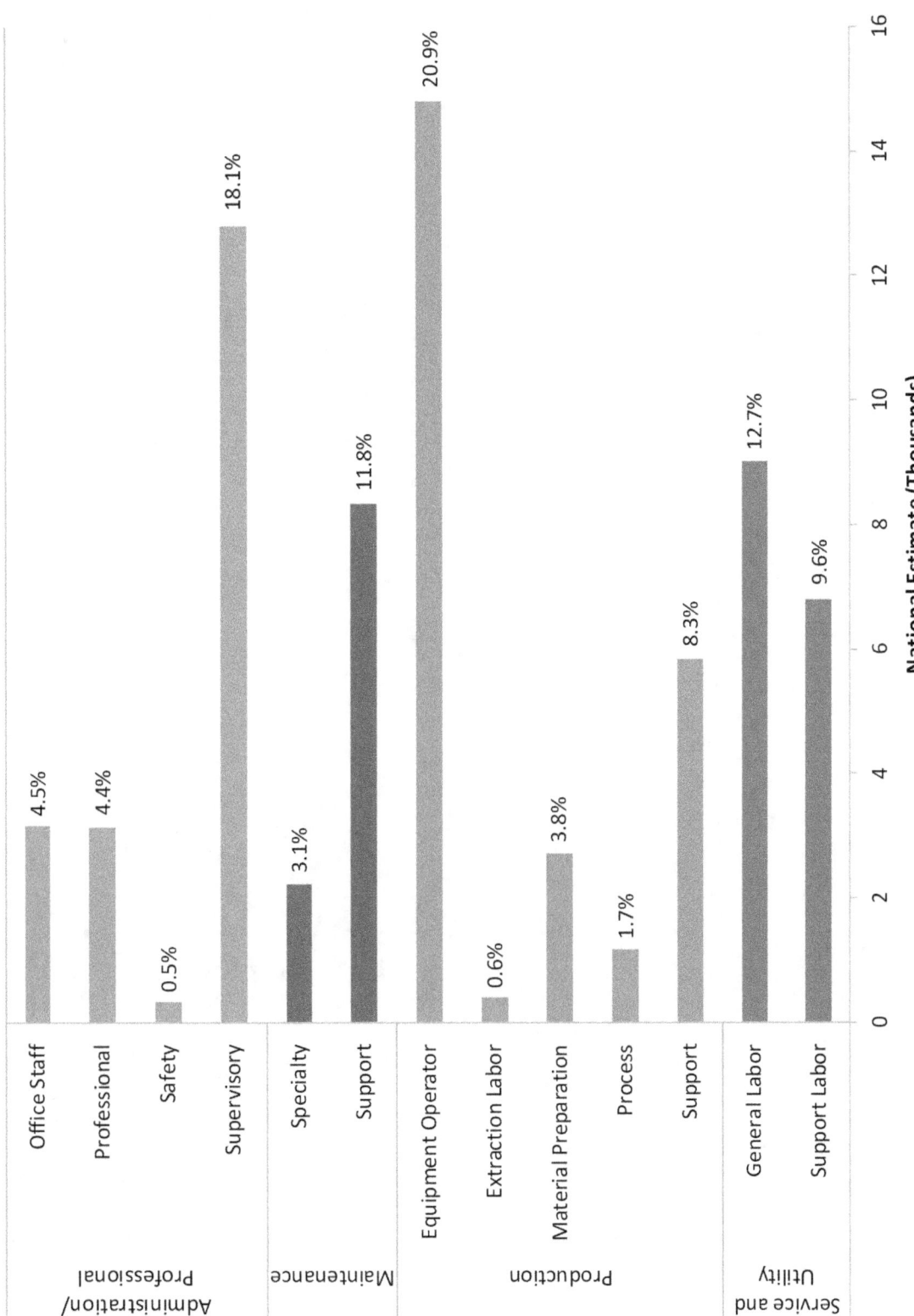

Figure 29. Occupational Categories of Employees at Stone Mines.

Employee Statistics for Sand and Gravel Mines

Summary of Employee Statistics for Sand and Gravel Mines

The demographic and occupational characteristics of employees in the U.S. sand and gravel mining industry are presented in Tables 51 and 52 and Figures 30–32. The weighted survey estimate for gender indicates that the workforce is composed predominately of male employees (92.1 percent). The majority of sand and gravel mine employees are White (94.1 percent), followed by Black or African American (4.0 percent). Almost 18 percent of these employees have an ethnicity of Hispanic or Latino. An estimated 59.9 percent are high school graduates, with another 20.7 percent having a level of education beyond high school. A review of the weighted estimates indicates that the average sand and gravel mine worker is 44.0 years of age and has worked in mining for 10.3 years, with 7.4 years at the current mine, and 7.4 years in his/her job title. The national estimate for the average number of hours worked per week is 46.1. The primary work location for an estimated 53.0 percent of sand and gravel mine employees is a "Surface Mine: Strip, Open Pit, or Quarry." An additional 14.8 percent of these employees work at a "Surface Mine: Other Surface Mining," while another 13.0 percent are employed in the "Surface Mine: Dredge" work location.

Tables 53, 54, 56, 57, and Figure 33 present the national estimates of the number of sand and gravel mine workers by four major occupational categories. (No estimates were calculated for Table 55: "Miscellaneous.") An estimated 9,445 (29.5 percent) are employed in the "Administration/Professional" category; 2,640 (8.3 percent) in the "Maintenance" category; 11,971 (37.5 percent) in the "Production" category; and 7,928 (24.7 percent) in the "Service and Utility" category.

Table 51. Demographic Characteristics of Employees at Sand and Gravel Mines

Demographic Characteristic	Survey Count	National Estimate	95% LCL	95% UCL	National Percent	95% LCL	95% UCL
Gender:							
Male	1,280	29,343	24,178	34,508	92.1	90.0	94.2
Female	109	2,531	1,607	3,456	7.9	5.8	10.0
Age (years)	1,326	44.0	43.0	45.1			
Highest level of education:							
Less than 9th grade	69	1,464	310	2,619	4.8	1.1	8.4
9th–12th grade (no diploma)	176	4,502	3,122	5,881	14.7	10.3	19.1
HS Graduate or Equivalent (GED)	817	18,394	14,222	22,566	59.9	54.0	65.9
Some College, Associate Degree, or Technical School	238	5,276	3,850	6,701	17.2	14.4	19.9
Bachelor's Degree or beyond	60	1,065	656	1,475	3.5	2.3	4.6
Ethnicity:							
Hispanic or Latino	286	5,154	2,850	7,458	17.5	9.3	25.6
Non-Hispanic or Non-Latino	1,027	24,345	18,596	30,093	82.5	74.4	90.7
Race:							
American Indian or Alaska Native	28	441	210	673	1.6	0.7	2.5
Asian	4	DSU	DSU	DSU	DSU	DSU	DSU
Black or African American	58	1,109	277	1,940	4.0	1.6	6.3
Native Hawaiian or Other Pacific Islander	2	DSU	DSU	DSU	DSU	DSU	DSU
White	1,066	26,151	20,354	31,948	94.1	92.1	96.1

Abbreviation: DSU, data suppressed.

Table 52. Occupational Characteristics of Employees at Sand and Gravel Mines

Occupational Characteristic	Survey Count	National Estimate	95% LCL	95% UCL	National Percent	95% LCL	95% UCL
Hours worked (per week)	1,220	46.1	43.6	48.5			
Experience:							
Experience in this Job Title (years)	1,350	7.4	6.2	8.6			
Experience at this Mine (years)	1,335	7.4	6.2	8.7			
Total Mining Experience (years)	1,352	10.3	9.5	11.2			
Primary Work Location:							
Surface Mine: Strip, Open Pit, or Quarry	678	17,029	12,145	21,913	53.0	42.3	63.7
Surface Mine: Dredge	171	4,190	2,281	6,100	13.0	7.8	18.3
Surface Mine: Other Surface Mining (Metal/Nonmetal Only)	244	4,740	2,462	7,018	14.8	7.3	22.2
Independent Shops or Yards	7	153	27	278	0.5	0.1	0.8
Mill Operations, Preparation Plants, or Breakers	129	2,305	1,078	3,532	7.2	3.3	11.0
Office	162	3,701	2,470	4,932	11.5	8.6	14.4

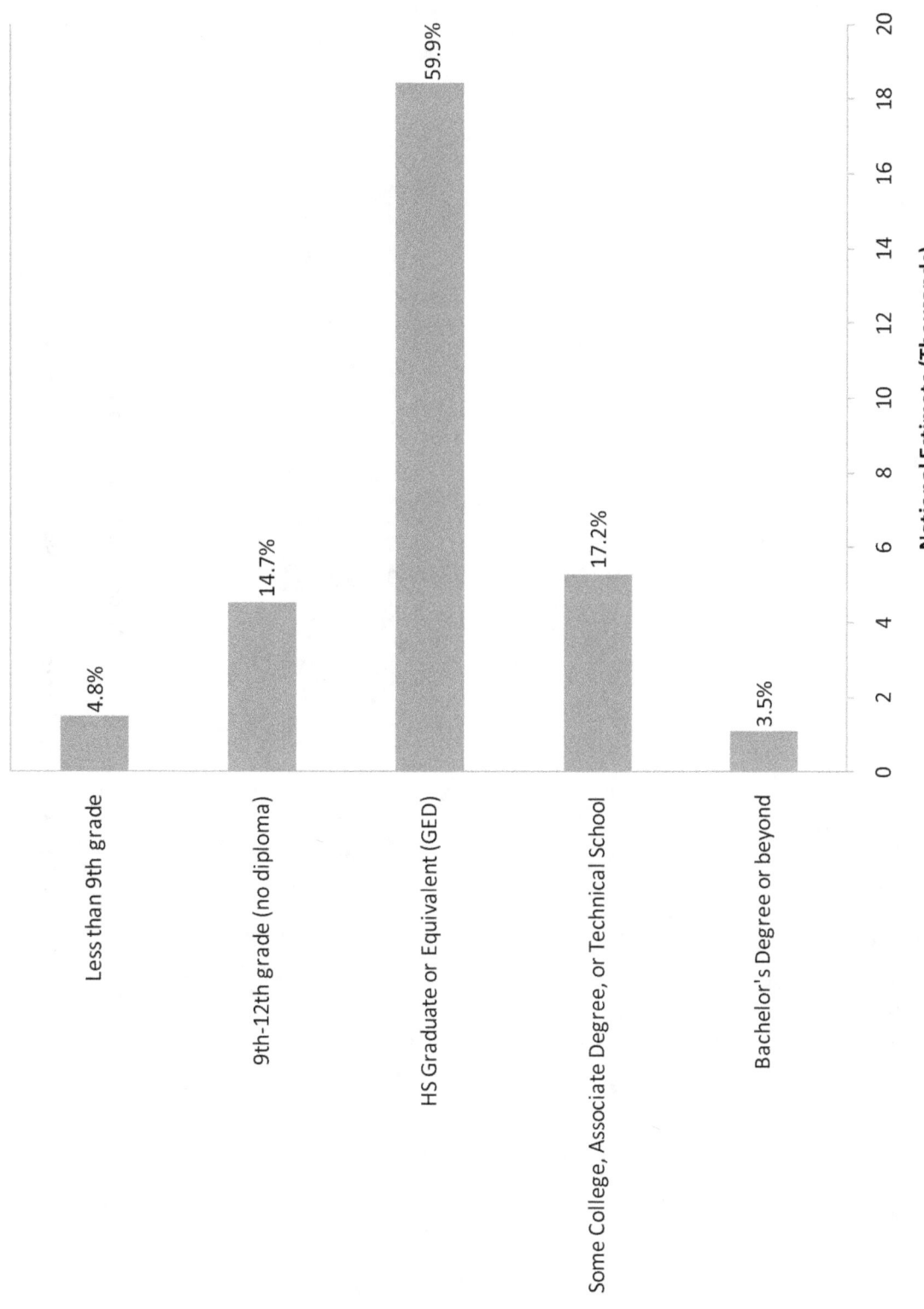

Figure 30. Education Level of Employees at Sand and Gravel Mines.

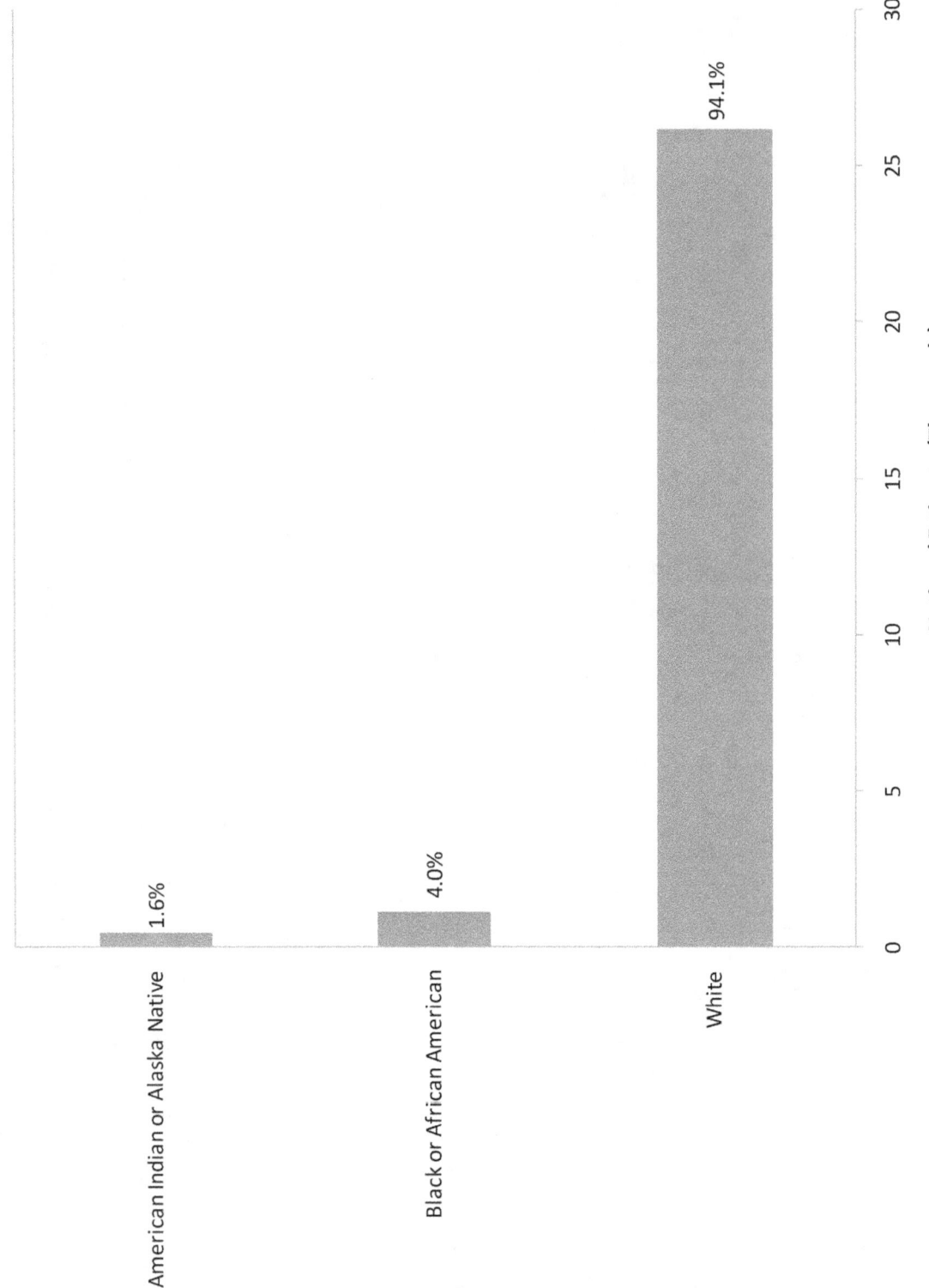

Figure 31. Race of Employees at Sand and Gravel Mines.

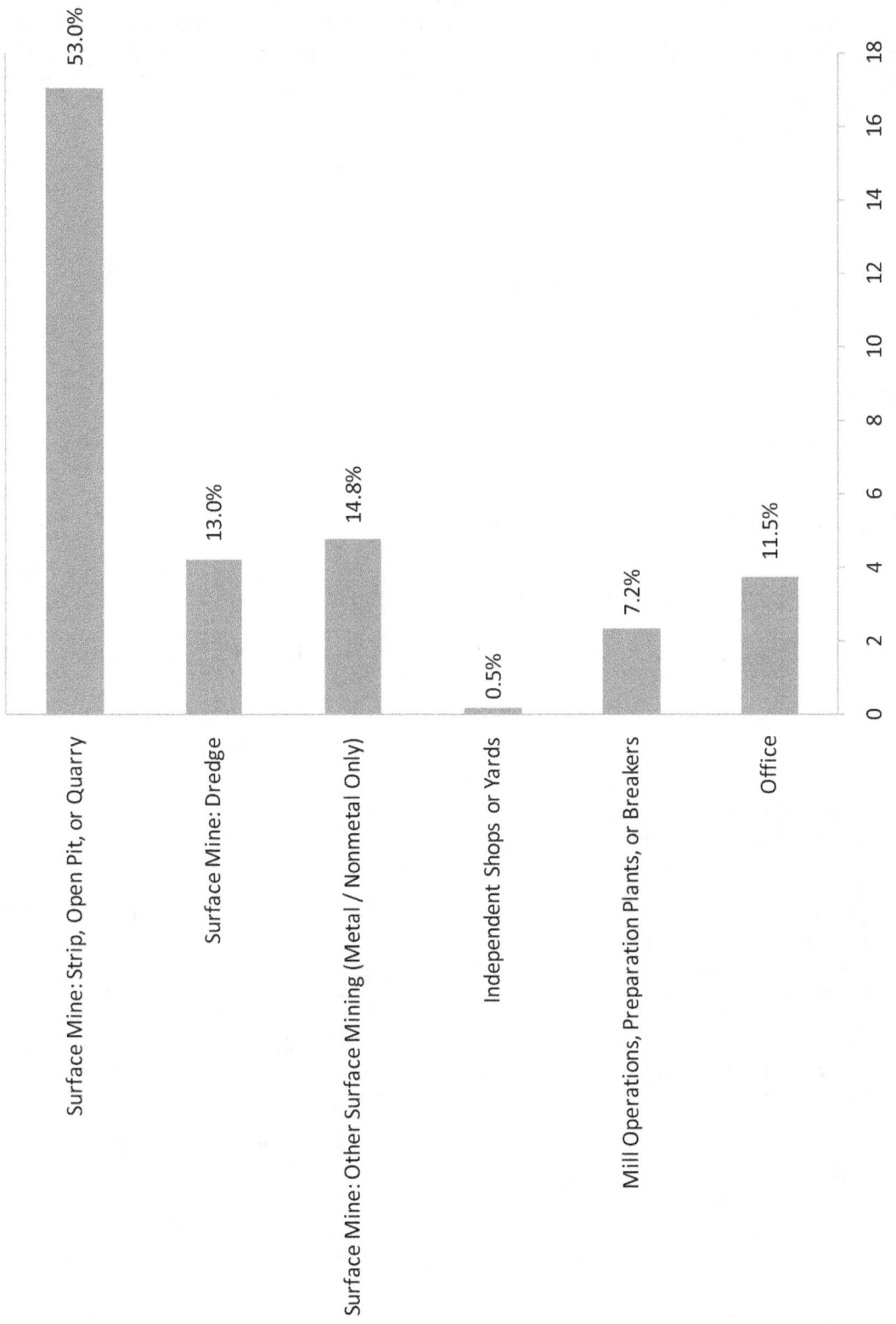

Figure 32. Primary Work Location of Employees at Sand and Gravel Mines.

Table 53. Estimated Number of Administration/Professional Employees at Sand and Gravel Mines

Occupation by Category	Survey Count	National Estimate	95% LCL	95% UCL
ADMINISTRATION/PROFESSIONAL	**398**	**9,445**	**6,998**	**11,892**
<u>*Office Staff*</u>	<u>65</u>	<u>1,512</u>	<u>745</u>	<u>2,279</u>
Administrative Staff	43	1,098	401	1,795
Administration				
Administrative Assistant				
Clerk				
Customer Service				
Office Clerk				
Office Staff				
Plant Clerk				
Receptionist				
Secretary				
Business	17	350	88	612
Accounting				
Bookkeeper				
Payroll				
Purchasing				
Sales				
Security	3	DSU	DSU	DSU
Guard				
Union Representative	2	DSU	DSU	DSU
<u>*Professional*</u>	<u>33</u>	<u>596</u>	<u>185</u>	<u>1,007</u>
Engineer	7	61	0	124
Engineer				
(Electrical/Mining/Ventilation)				
Engineer, not otherwise specified				
Environmental Engineer				
Plant Engineer				
Non-engineer	12	302	0	671
Environmental Specialist				
Operating Engineer				
Production Scheduler				
Technician	14	234	74	393
Sampler/Lab Technician				
Technician				
<u>Safety</u>	<u>10</u>	<u>165</u>	<u>68</u>	<u>262</u>
Safety				
Safety Manager				
Safety Supervisor				

Table 53. Estimated Number of Administration/Professional Employees at Sand and Gravel Mines (continued)

Occupation by Category	Survey Count	National Estimate	95% LCL	95% UCL
<u>Supervisory</u>	<u>290</u>	<u>7,172</u>	<u>5,188</u>	<u>9,156</u>
Executive	*8*	*348*	*147*	*550*
General Manager				
Mine Owner				
President				
Vice President				
Foreman	*86*	*1,868*	*1,352*	*2,383*
Foreman				
Foreman/Shift Boss				
Lead Man				
Maintenance Foreman				
Maintenance Lead Man				
Shop Foreman				
Plant Foreman				
Superintendent				
Manager	*60*	*1,485*	*848*	*2,121*
Area Manager				
Assistant Manager				
Dredge Manager				
Dry Plant Manager				
Equipment Manager				
Manager				
Office Manager				
Operations Manager				
Plant Manager				
Production Manager				
Purchasing Manager				
Quarry Manager				
Sales Manager				
Shift Manager				
Shop Manager				
Supervisor	*136*	*3,471*	*2,234*	*4,708*
Backhoe Supervisor				
Dozer Supervisor				
Lab Supervisor				
Maintenance Supervisor				
Mine Operator				
Plant Operator				
Plant Supervisor				
Production Supervisor				
Quarry Supervisor				
Shift Supervisor				
Supervisor				

Abbreviation: DSU, data suppressed.

Table 54. Estimated Number of Maintenance Employees at Sand and Gravel Mines

Occupation by Category	Survey Count	National Estimate	95% LCL	95% UCL
MAINTENANCE	176	2,640	2,135	3,145
<u>Specialty</u>	<u>17</u>	<u>274</u>	<u>88</u>	<u>460</u>
Electrician	6	78	0	184
Electrician/Wireman				
Maintenance Electrician				
Welder	11	196	40	351
Certified Welder				
Repair/Welder				
Welder				
Welder/Mechanic				
<u>Support</u>	<u>159</u>	<u>2,365</u>	<u>1,910</u>	<u>2,820</u>
Maintenance	56	803	583	1,022
Electrical Maintenance				
Fixed Maintenance				
Greaser/Oiler				
Maintenance				
Maintenance Planner				
Plant Maintenance				
Production/Process Maintenance				
Truck Maintenance				
Mechanic	69	1,125	775	1,476
Aggregate Mechanic				
Equipment Mechanic				
Maintenance Mechanic				
Mechanic				
Mechanic Helper				
Mechanic/Welder				
Mobile Equipment Mechanic				
Mobile Maintenance Mechanic				
Mobile Mechanic				
Plant Mechanic				
Repairman	34	437	108	766
Automotive Repairman				
Heavy Duty Repairman				
Plant Repairman				
Repairman				
Skilled Repairman				

Table 55. Number of Miscellaneous Employees at Sand and Gravel Mines

Occupation by Category	Survey Count
MISCELLANEOUS	**6**
Trainee	1
Unknown	5

Table 56. Estimated Number of Production Employees at Sand and Gravel Mines

Occupation by Category	Survey Count	National Estimate	95% LCL	95% UCL
PRODUCTION	**506**	**11,971**	**7,813**	**16,130**
Equipment Operator	*311*	*7,118*	*4,308*	*9,927*
Dragline Operator	*9*	*194*	*21*	*367*
Equipment Operator	**194**	**4,530**	**2,398**	**6,663**
Backhoe Operator				
Bobcat Operator				
Bulldozer Operator				
Dredge Operator				
Equipment Operator				
Front End Loader Operator				
Grader Operator				
Heavy Equipment Operator				
Highlift Operator				
Hopper Operator				
Mobile Equipment Operator				
Mucking Machine Operator				
Rotary Bucket Excavator Operator				
Stripping Operator				
Tower Operator				
Track Hoe				
Tractor Operator				
Hoist	*16*	*111*	*0*	*368*
Hoist Engineer				
Material Mover	*85*	*2,174*	*1,024*	*3,323*
Haul Truck Operator/Driver				
Pit Truck Driver				
Rock Truck Driver				
Stock Truck/Stock Pile Driver				
Truck Driver				
Water Truck Operator				
Operator	*4*	*DSU*	*DSU*	*DSU*
Motorman				
Shovel Operator	*3*	*DSU*	*DSU*	*DSU*

Table 56. Estimated Number of Production Employees at Sand and Gravel Mines (continued)

Occupation by Category	Survey Count	National Estimate	95% LCL	95% UCL
Material Preparation	<u>43</u>	<u>1,110</u>	<u>359</u>	<u>1,861</u>
Crusher	26	973	225	1,722
Crusher Operator/Pan Feeder Operator				
Crusher Plant Operator				
Cutter	11	27	0	75
Splitter				
Mill	6	110	0	265
Mill Operator (ball/pebble/rod)				
Process	<u>24</u>	<u>511</u>	<u>224</u>	<u>798</u>
Dry Processing	12	191	37	344
Dry Plant/Process Operator				
Dryer Operator				
Fluid Bed Dryer Operator				
Separation	2	DSU	DSU	DSU
Pug Operator/Mixer Tender				
Slurry Operator				
Wash Process	9	267	15	519
Wash Operator				
Wet Process	1	DSU	DSU	DSU
Wet Plant Operator				
Support	<u>128</u>	<u>3,233</u>	<u>2,071</u>	<u>4,395</u>
Drill Operator	2	DSU	DSU	DSU
Explosives	2	DSU	DSU	DSU
Blaster				
Other	107	2,806	1,626	3,987
Dispatcher				
Operator, not otherwise specified				
Production Operator				
Quality Control	17	393	70	716
Quality Control/Quality Assurance				

Abbreviation: DSU, data suppressed.

Table 57. Estimated Number of Service and Utility Employees at Sand and Gravel Mines

Occupation by Category	Survey Count	National Estimate	95% LCL	95% UCL
SERVICE and UTILITY	**306**	**7,928**	**6,032**	**9,824**
<u>General Labor</u>	<u>150</u>	<u>3,470</u>	<u>2,119</u>	<u>4,822</u>
Cleaners	*1*	*DSU*	*DSU*	*DSU*
Cleanup Man				
Laborer	*81*	*1,916*	*876*	*2,957*
Ground Hand				
Ground Man				
Laborer				
Plant Helper				
Plant Man				
Root Picker				
Stick Picker				
Material Handling	*22*	*380*	*0*	*772*
Bagger/Bagging Operations Worker				
Mudpicker				
Reclaim Operator				
Storeroom				
Sweeper Operator				
Tradesman	*3*	*DSU*	*DSU*	*DSU*
Apprentice/Journeyman				
Weighman	*43*	*1,104*	*577*	*1,631*
Scale Clerk/Operator				
Weighmaster				
<u>Support Labor</u>	<u>156</u>	<u>4,458</u>	<u>3,352</u>	<u>5,563</u>
Barge Operations	*4*	*DSU*	*DSU*	*DSU*
Barge Attendant/Boat Operator				
Deck Hand				
Conveyor Operator	*4*	*DSU*	*DSU*	*DSU*
Belt Cleaner/Conveyor Man				
Distribution	*1*	*DSU*	*DSU*	*DSU*
Packaging Operations Worker				
Loading	*123*	*3,919*	*2,766*	*5,072*
Bulk Loader				
Loader Operator				
Loadout Operator				
Operator/Loader				
Plant Loader Operator				
Rail Loader Operator				
Shipping Loader				
Stock Loader/Piler				
Yard Loader Operator				

Table 57. Estimated Number of Service and Utility Employees at Sand and Gravel Mines (continued)

Occupation by Category	Survey Count	National Estimate	95% LCL	95% UCL
Pumper	1	*DSU*	*DSU*	*DSU*
Gravel Pumper				
Supplies	2	*DSU*	*DSU*	*DSU*
Parts				
Utility	21	361	200	523
Equipment Utility				
Pit Utility Person				
Plant Utility				
Utility Beltline				
Utility Man				
Wet Utility				

Abbreviation: DSU, data suppressed.

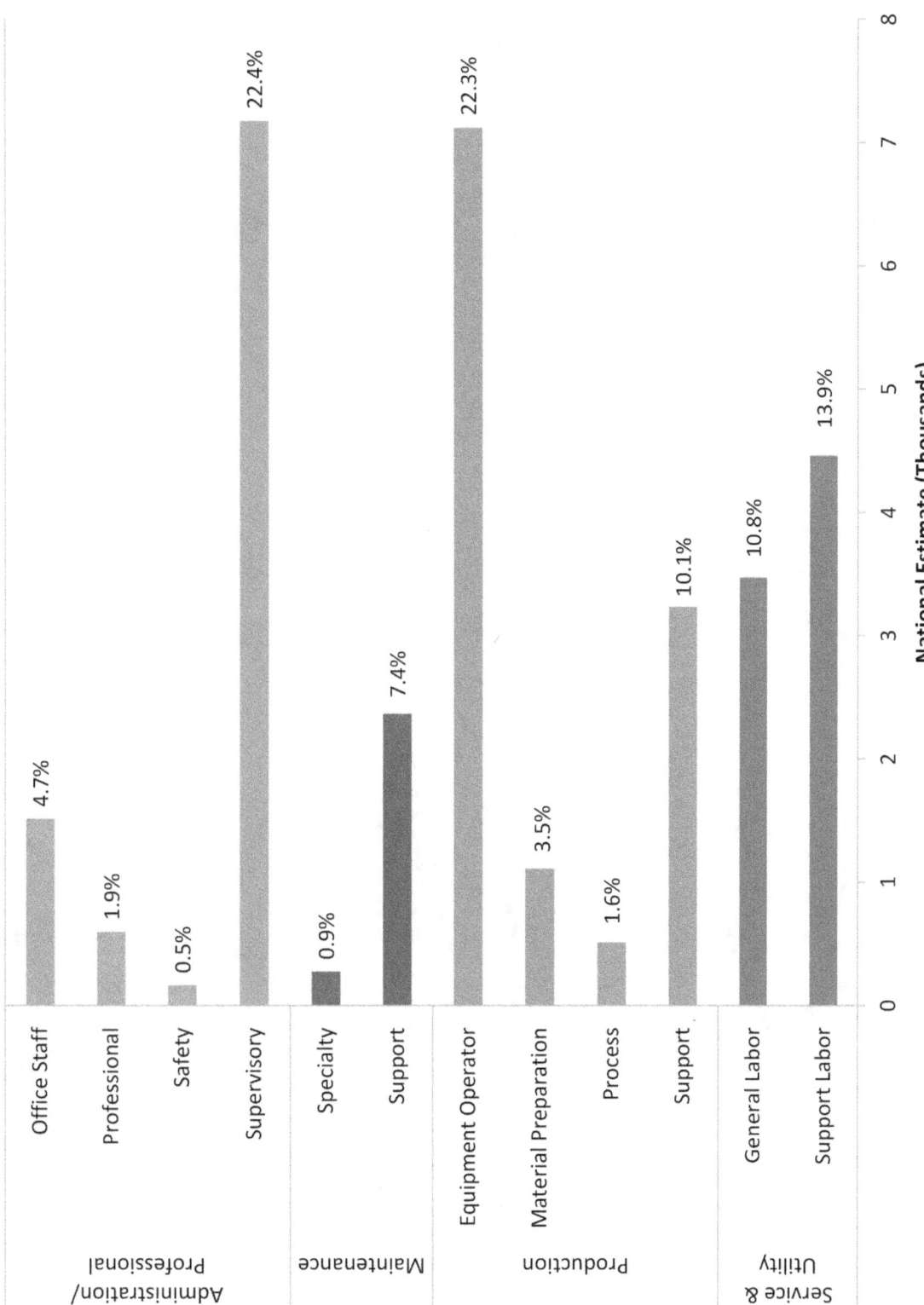

Figure 33. Occupational Categories of Employees at Sand and Gravel Mines.

147

Acknowledgements

The authors extend their appreciation to the reviewers for their thoughtful and insightful comments. We acknowledge the contributions of: Jefferey L. Burgess, M.D., M.P.H., University of Arizona, Jeffery H. Kravitz, Ph.D., Mine Safety and Health Administration, Elaine N. Rubinstein, Ph.D., University of Pittsburgh, John P. Sestito and William K. Sieber, National Institute for Occupational Safety and Health, and Bruce Watzman, National Mining Association.

Additionally, we appreciate the contributions of our numerous colleagues at the Office of Mine Safety and Health Research. In particular, Audrey Podlesny for preparing the maps, Michael J. Brnich, Jr., Albert H. Cook, Nathan T. Lowe, Mary Ellen Nelson, Robert H. Peters, and Charles Vaught (retired) for their invaluable advice regarding the coding of the mining industry job titles, Joseph Schall and Candace A. Wolf for their editorial assistance, Donna Opfer and Brittney Warnick for their formatting assistance, Genny Lohr for her work on 508 Compliance, and David Beshero for the cover design.

Finally, the authors gratefully acknowledge the survey respondents in the coal, metal, nonmetal, stone, and sand and gravel mining sectors. Thank you for your time and effort in completing this survey.

References

American Geological Institute [1997]. Dictionary of mining, mineral, and related terms. Alexandria, VA: American Geological Institute.

BLS [2010]. Standard occupational classification. Washington, DC: U.S. Department of Labor, Bureau of Labor Statistics, Standard Occupational Classifications [http://www.bls.gov/SOC/].

Cecala A and Thimons E [1992]. Tips for reducing dust from secondary sources during bagging. Powder and Bulk Eng. 7(5):77.

Cochran WG [1977]. Sampling techniques, 3rd ed. New York: John Wiley & Sons, pp. 127–131.

Dictionary.com [2011]. [http://www.dictionary.reference.com].

DOT [2003]. Dictionary of occupational titles [http://www.occupationalinfo.org].

Infomine Inc. [2010]. Dictionary of mining and mineral terms [http://www.infomine.com/dictionary/].

Merriam-Webster [2011]. [http://www.merriam-webster.com/].

MSHA [2007]. Part 50 data user's handbook. Denver: U.S. Department of Labor, Mine Safety and Health Administration, Office of Injury and Employment Information.

MSHA [2011]. Program information bulletin No. P11-08. U.S. Department of Labor, Mine Safety and Health Administration [http://www.msha.gov/regs/complian/PIB/2011/pib11-08.asp].

NCHS [2002]. Statistical notes number 24. Healthy people 2010 criteria for data suppression. By Klein RJ, Proctor SE, Boudreault MA, Turczyn KM. Hyattsville, MD: U.S. Department of Health and Human Services, Centers for Disease Control and Prevention, National Center for Health Statistics. DHHS Publication No. (PHS) 2002-1237 2-0424.

NCHS [2004]. NCHS staff manual on confidentiality. U.S. Department of Health and Human Services, Centers for Disease Control and Prevention, National Center for Health Statistics.

Thrush P [1968]. A dictionary of mining, mineral, and related terms. Washington, DC: U.S. Department of the Interior, Bureau of Mines.

Vaught C (Retired) [2008]. Glossary of mining terms. Private email message to Linda McWilliams (LMcWilliams@cdc.gov), March 28.

Wang Z, Waldron W [2010]. Using the SAS® survey procedures for subpopulation analysis with jackknife repeated replication methods in SAS 9.2. In: Proceedings of the SAS Global Forum, pp. 1–9.

Appendices

Appendix A. Questionnaire Booklet

Form Approved
OMB NO. 0920-0754
Exp. Date 10/31/2010

National Survey of the Mining Population Questionnaire

SAFER • HEALTHIER • PEOPLE™

**Centers for Disease Control and Prevention
National Institute for Occupational Safety and Health
Pittsburgh Research Laboratory
P.O. Box 18070
Pittsburgh, Pennsylvania 15236**

Public reporting burden of this collection of information is estimated to average 120 minutes per response, including the time for reviewing instructions, searching existing data sources, gathering and maintaining the data needed, and completing and reviewing the collection of information. An agency may not conduct or sponsor, and a person is not required to respond to a collection of information unless it displays a currently valid OMB control number. Send comments regarding this burden estimate or any other aspect of this collection of information, including suggestions for reducing this burden to CDC/ATSDR Reports Clearance Officer; 1600 Clifton Road NE, MS E-11, Atlanta, Georgia 30333; ATTN: PRA (0920-0633).

Mine ID Number:
«MineIDNumber»

Mine Name:
«MineName»

Reporting Week:
«ReportingWeek»

QUESTIONNAIRE OVERVIEW

This questionnaire contains five parts:

- Mine Questions — Pages 1-13
- Employee Selection Instructions — Page 14
- Employee Questions Instructions — Pages 15-16
- Employee Questions — Pages 17-18
- Final Questions and Comments — Pages 19-20

Items of Special Importance:

1. All responses you give should be for the specific Mine ID and name shown in the box above. Some items in the questionnaire are for a specific one-week period called the REPORTING WEEK, which is your payroll week that includes the date shown in the box above.

2. You have the option of completing either this survey questionnaire booklet or an Internet web-based survey questionnaire. The contents of both versions of the survey questionnaire are the same. Instructions to access the web-based questionnaire (www.miningsurvey.org) are attached to the cover letter included in the survey mailing.

3. If you have a question regarding your REPORTING WEEK, how to access the web-based questionnaire, or if you need assistance in completing any of the items, please call 1-888-814-4707. This is the toll-free number for Westat, the survey contractor.

4. Use the Comments section (Item F8 on Page 20) to explain any responses or situations unique to your mine.

MINE QUESTIONS

TRAINING

The first series of questions asks about miner training. This includes both **annual miner refresher training** and **new miner training**.

M1. In the past 12 months, did this mining operation use its *employees* to conduct. **[Please check "Yes" or "No" for each question a, b, and c below.]**

		Yes	No
a.	annual miner refresher training?	☐	☐
b.	training for newly hired *inexperienced* miners?	☐	☐
c.	training for newly hired *experienced* miners?	☐	☐

M2. In the past 12 months, did this mining operation use an *outside trainer* to conduct **annual miner refresher training**?

☐ Yes → Go to Question M3
☐ No → Go to Question M4 (next page)

M3. [IF YES TO Question M2]: What type of *outside trainer* did you use? **[Please check ALL that apply.]**

☐ Contract trainer
☐ State grantee
☐ Other (**Please specify**):
_____
_____

TRAINING (continued)

M4. In the past 12 months, did this mining operation use an *outside trainer* to conduct training for newly hired *inexperienced* miners?

☐ Yes → Go to Question M5

☐ No → Go to Question M6

M5. [IF YES TO Question M4]: What type of *outside trainer* did you use? [Please check ALL that apply.]

☐ Contract trainer
☐ State grantee
☐ Other (Please specify):

_____

_____

M6. In the past 12 months, did this mining operation use an *outside trainer* to conduct training for newly hired *experienced* miners?

☐ Yes → Go to Question M7

☐ No → Go to Question M8

M7. [IF YES TO Question M6]: What type of *outside trainer* did you use? [Please check ALL that apply.]

☐ Contract trainer
☐ State grantee
☐ Other (Please specify):

_____

_____

M8. How frequently are periodic safety meetings (e.g., "toolbox talks"), for employees engaged in mining operations, conducted at this mine? [Please check one.]

☐ Less than once a year
☐ Annually
☐ Less than once a month
☐ Once a month
☐ Once every 2 weeks
☐ Once a week
☐ Several times a week
☐ Daily

M9. When conducting employee safety training and retraining, which of the following training materials and methods are used as part of your training program? [Please check ALL that apply.]

☐ Lectures
☐ Written materials
☐ Videos
☐ Self-guided interactive computer programs
☐ Demonstrations
☐ Hands-on training exercises
☐ Group exercises (role playing, games, problem solving, etc.)
☐ Classroom simulations (e.g., virtual reality)
☐ Worksite simulations
☐ Narrative story telling
☐ Other (Please specify):

_____

_____

OTHER LANGUAGES

The next series of questions asks about the use of languages other than English.

M10. Approximately what percentage of employees currently working at the mine use a language other than English to communicate?

_____%

M11. Does this mining operation currently provide training materials, signs, or other written materials in a language other than English?

☐ Yes → Go to Question M12
☐ No → Go to Question M14

M12. [IF YES TO Question M11]: What language(s) is/are provided? [Please check ALL that apply.]

☐ Spanish
☐ Other (Please specify):
_____
_____

M13. Would it be helpful to have training materials, signs, or written materials in any other languages, *in addition* to those already provided by your mining operation?

☐ Yes → Go to Question M15
☐ No → Go to Work Schedules Section (next page)

M14. Would it be helpful to have training materials, signs, or other written materials in language(s) other than English?

☐ Yes → Go to Question M15
☐ No → Go to Work Schedules Section (next page)

M15. [IF YES TO Question M13 or M14]: Which languages? [Please check ALL that apply.]

☐ Spanish
☐ Other (Please specify):
_____
_____

WORK SCHEDULES

The next series of questions asks about how the mine schedules work for the following types of mine operator employees:

- **Production Workers** are 'face workers' and others who work extracting coal/ore/stone.
- **Production Support Workers** are those who aid and maintain production (e.g., by cleaning or moving belts, maintaining ventilation, delivering supplies, repairing equipment, etc. Office workers are also counted here).
- **Preparation Plant/Mill Workers** are those who operate or perform support activities in a preparation plant or mill.

We suggest, for this section and the next, that you first respond to all questions in Column A for Production Workers, and then go back to complete them in Column B for Production Support Workers, followed by the Column C items for Preparation Plant/Mill Workers.

WORK SCHEDULES	A. Production Workers	B. Production Support Workers	C. Preparation Plant/ Mill Workers
	☐ **CHECK** If this mine does not have any Production Workers and leave this column blank. If Box is **NOT CHECKED**, continue with this column.	☐ **CHECK** if this mine does not have any Production Support Workers (and no office workers), then leave this column blank. If Box is **NOT CHECKED**, continue with this column.	☐ **CHECK** if this mine does not have any Preparation Plant/Mill Workers and leave this column blank. If Box Is **NOT CHECKED**, continue with this column.
M16.a. On average, how many days per week are these workers *scheduled* to work?	|__| Scheduled days per week	|__| Scheduled days per week	|__| Scheduled days per week
b. On average, how many hours per day are these workers *scheduled* to work?	|__|__| Scheduled hours per day	|__|__| Scheduled hours per day	|__|__| Scheduled hours per day

WORK SCHEDULES (continued)

WORK SCHEDULES	A. Production Workers	B. Production Support Workers	C. Preparation Plant/ Mill Workers
M16.c. During the REPORTING WEEK (which includes the date shown in the box on Page 1), what was the average number of hours per week these workers *actually* worked (including overtime)?	|_|_|_| Actual work hours during REPORTING WEEK	|_|_|_| Actual work hours during REPORTING WEEK	|_|_|_| Actual work hours during REPORTING WEEK
d. Do work crews generally change shifts at the active mining site (e.g., the face or long wall - also known as a 'hot seat' change)?	☐ Yes ☐ No	☐ Yes ☐ No	QUESTIONS M16.d & e. NOT APPLICABLE FOR PREPARATION PLANT MILL WORKERS GO TO SHIFT WORK SECTION (Next Page)
e. On average, how much time per shift do workers spend traveling to and from the active mining site on-shift (while being paid)?	|_| |_|_| Hours Minutes round trip, per shift GO TO COLUMN B	|_| |_|_| Hours Minutes round trip, per shift GO TO COLUMN C	

SHIFT WORK

For the next series of questions, assume that the:

- **Day** shift begins in the morning hours (e.g., 6 a.m., 7 a.m., or 8 a.m.)
- **Afternoon** shift begins in the afternoon hours (e.g., 2 p.m. or 3 p.m.)
- **Night** or **Midnight** shift begins in the late evening hours (e.g., 11 p.m. or 12 a.m.)

SHIFT WORK	A. Production Workers	B. Production Support Workers	C. Preparation Plant/ Mill Workers
M17. Typically how many shifts per day does the mine operate for these workers?	☐ **CHECK** if this mine does not have any Production Workers and leave this column blank. If Box is **NOT CHECKED**, continue with this column. |_| Shifts per day	☐ **CHECK** if this mine does not have any Production Support Workers (and no office workers), then leave this column blank. If Box is **NOT CHECKED**, continue with this column. |_| Shifts per day	☐ **CHECK** if this mine does not have any Preparation Plant/Mill Workers and leave this column blank. If Box is **NOT CHECKED**, continue with this column. |_| Shifts per day
M18. Do they work rotating shifts?	☐ Yes → GO TO QUESTION M19 ☐ No → GO TO QUESTION M21 (Next Page)	☐ Yes → GO TO QUESTION M19 ☐ No → GO TO QUESTION M21 (Next Page)	☐ Yes → GO TO QUESTION M19 ☐ No → GO TO QUESTION M21 (Next Page)
M19. [IF YES TO QUESTION M18]: How frequently do these workers change their assigned shift?	☐ Weekly ☐ Twice a Month ☐ Monthly ☐ Other (specify): _____	☐ Weekly ☐ Twice a Month ☐ Monthly ☐ Other (specify): _____	☐ Weekly ☐ Twice a Month ☐ Monthly ☐ Other (specify): _____

SHIFT WORK (continued)

SHIFT WORK	A. Production Workers	B. Production Support Workers	C. Preparation Plant/ Mill Workers
M20. Do they rotate shifts clockwise or counterclockwise? Note that *Clockwise* is **day→afternoon→night** *Counterclockwise* is **night→afternoon→day**	☐ Clockwise ☐ Counterclockwise ☐ Other **(specify):** _____	☐ Clockwise ☐ Counterclockwise ☐ Other **(specify):** _____	☐ Clockwise ☐ Counterclockwise ☐ Other **(specify):** _____
M21. Are there any regularly scheduled *unique* work shifts that do not fit into the previous descriptions (e.g., a shift of three 12-hour days on Friday, Saturday, and Sunday, known as an "alternative work schedule" or "Weekend Warrior" shift)?	☐ Yes → GO TO QUESTION M22 ☐ No → GO TO COLUMN B	☐ Yes → GO TO QUESTION M22 ☐ No → GO TO COLUMN C	☐ Yes → GO TO QUESTION M22 ☐ No → GO TO NEXT PAGE
M22. [IF YES TO QUESTION M21]: Please **either**: a. describe this shift. If you need additional space, use the 'comments' section (Item F8) on Page 20; **or:** b. send us an example of your mine's shift schedule(s) and check the appropriate box(es).	_____ _____ _____ _____ _____ _____ _____ ☐ Schedule enclosed	_____ _____ _____ _____ _____ _____ _____ ☐ Schedule enclosed	_____ _____ _____ _____ _____ _____ _____ ☐ Schedule enclosed

INDEPENDENT CONTRACTOR EMPLOYEES

The next series of questions asks about the mine's use of independent contractor employees for various activities. Take special note of these two definitions:

- **Independent contractor** means "any person, partnership, corporation, firm, association, subsidiary of a corporation, or other organization that contracts to perform services or construction of a mine."

- **REPORTING WEEK** is your specific 7-day payroll period that includes the date shown in the box on Page 1. The number of independent contractors you report should be for that week only.

M23. In the REPORTING WEEK, did this mining operation use independent contractor employees to do . . .	M24. How many independent contractor employees did you use for this activity during the REPORTING WEEK?	M25. How many total hours did independent contractor employees work in this activity during the REPORTING WEEK?
a. Mine development, including shaft and slope sinking, or "driving a decline"? ☐ Yes ☐ No	a. _____ → # of Contractor employees	a. _____ Contractor hours
b. Construction or reconstruction of mine facilities, including building or rebuilding preparation plants and mining equipment, maintenance, and building additions to existing facilities? ☐ Yes ☐ No	b. _____ → # of Contractor employees	b. _____ Contractor hours

INDEPENDENT CONTRACTOR EMPLOYEES (continued)

M23. In the REPORTING WEEK, did this mining operation use independent contractor employees to do ...	M24. How many independent contractor employees did you use for this activity during the REPORTING WEEK?	M25. How many total hours did independent contractor employees work in this activity during the REPORTING WEEK?
c. Demolition of mine facilities? ☐ Yes ☐ No	c. _____ → # of Contractor employees	c. _____ Contractor hours
d. Construction of dams? ☐ Yes ☐ No	d. _____ → # of Contractor employees	d. _____ Contractor hours
e. Excavation or earthmoving activities involving mobile equipment? ☐ Yes ☐ No	e. _____ → # of Contractor employees	e. _____ Contractor hours
f. Equipment installation, such as crushers and mills? ☐ Yes ☐ No	f. _____ → # of Contractor employees	f. _____ Contractor hours

INDEPENDENT CONTRACTOR EMPLOYEES (continued)

M23. In the REPORTING WEEK, did this mining operation use independent contractor employees to do . . .	**M24.** How many independent contractor employees did you use for this activity during the REPORTING WEEK?	**M25.** How many total hours did independent contractor employees work in this activity during the REPORTING WEEK?
g. Equipment service or repair of equipment on mine property for a period exceeding 5 consecutive days at a particular mine? ☐ Yes ☐ No	g. _____ ➔ # of Contractor employees	g. _____ Contractor hours
h. Material handling such as hauling of coal, ore, or refuse within mine property? (Only include material handling conducted primarily on mine property.) ☐ Yes ☐ No	h. _____ ➔ # of Contractor employees	h. _____ Contractor hours
i. Drilling and blasting? ☐ Yes ☐ No	i. _____ ➔ # of Contractor employees	i. _____ Contractor hours

INDEPENDENT CONTRACTOR EMPLOYEES (continued)

M23. In the REPORTING WEEK, did this mining operation use independent contractor employees to do...	**M24.** How many independent contractor employees did you use for this activity during the REPORTING WEEK?	**M25.** How many total hours did independent contractor employees work in this activity during the REPORTING WEEK?
j. Production support work (belt moves, building stoppings, installing roof support, moving a longwall, relocating a large piece of mining equipment (including dismantling and reassembly), surveying, engineering work, etc.)? ☐ Yes ⟶ ☐ No	j. _____ → # of Contractor employees	j. _____ Contractor hours
k. Mineral extraction? ☐ Yes ☐ No	k. _____ → # of Contractor employees	k. _____ Contractor hours
l. Any other types of work? ☐ Yes ☐ No → GO TO NEXT PAGE	l. _____ → # of Contractor employees	l. _____ Contractor hours

Please describe this activity:

_____
_____
_____
_____

SAFETY, COMMUNICATION, AND RESCUE MEASURES

M26. Which of the following types of communication devices and systems does this mine currently use? **[Please check ALL that apply.]**

- ☐ Dedicated telephones
- ☐ Mine page phones
- ☐ Trolley phones
- ☐ Shaft or hoist phones
- ☐ Cell phones
- ☐ Voice Over Internet Protocol (VOIP) phones
- ☐ Handheld two-way radios
- ☐ Wireless paging devices
- ☐ Leaky feeder communications system (not running a PED)
- ☐ Personal emergency device (PED) cap lamp/pager
- ☐ Through-the-Earth (TTE) technology (other than a PED, e.g., Flexalert or TeleMag)
- ☐ Inductive coupled radios
- ☐ Ethernet
- ☐ TRACKER Tagging System
- ☐ Longwall face communication systems
- ☐ None of the above
- ☐ Other **(Please specify):**

_____
_____

M27. Which of the following personal locators, trackers, or other devices does this mine currently use to make miners more visible and to support escape in limited visibility situations? **[Please check ALL that apply.]**

- ☐ Electronic or computerized tagging or tracking systems/devices
- ☐ Tag boards (check-in/check-out)
- ☐ Reflective vests/clothing
- ☐ Chemical light sticks
- ☐ Lighted vests
- ☐ Laser lights/pointers
- ☐ Strobe lights
- ☐ None of the above
- ☐ Other **(Please specify):**

_____
_____

M28. Which of the following methods does this mine use for emergency incident early warning systems for miners? **[Please check ALL that apply.]**

- ☐ Stench gas
- ☐ Audible systems
- ☐ Visual systems (lights)
- ☐ Pager phones
- ☐ Telephones
- ☐ Messengers
- ☐ Electronic personal communication systems (e.g., PED)
- ☐ None of the above
- ☐ Other **(Please specify):**

_____
_____

SAFETY, COMMUNICATION, AND RESCUE MEASURES (continued)

M29. Does this mine have its own mine rescue team?

☐ Yes → **[IF YES]** How many individual members are assigned to the mine's rescue team?

_____

Record total members above and **Go to Question M30**

☐ No → **[IF NO]** Go to NOTE box in next column

M30. How frequently is team training conducted for the members of the mine rescue team? **[Please check one.]**
☐ Less than once a year
☐ Annually
☐ Less than once a month
☐ Once a month
☐ Once every 2 weeks
☐ Once a week
☐ Some other time interval **(Please specify):**

_____
_____

NOTE – The next two questions (M31 and M32) apply only to underground mines. Surface mine respondents should skip to the next section (Employee Selection Instructions).

M31. Which of the following types of emergency equipment or emergency supplies does this mine currently rely on for miner safety? **[Please check ALL that apply.]**
☐ Belt-worn self-contained-self-rescuers (SCSRs)
☐ Cached self-contained-self-rescuers (SCSRs)
☐ Filter self-rescuers (FSRs) (e.g., W65)
☐ Stationary emergency refuge chambers
☐ Mobile emergency refuge chambers
☐ Sealing materials
☐ Cached water/food supplies
☐ First aid kits
☐ Defibrillator
☐ None of the above
☐ Other **(Please specify):**

_____
_____

M32. Which of the following types of escapeway aids does this mine use? **[Please check ALL that apply.]**
☐ Lifelines
☐ Directional lifelines
☐ Signage
☐ Colored reflectors
☐ Lighting
☐ Strobe lights
☐ None of the above
☐ Other **(Please specify):**

_____
_____

Mine ID Number:	Reporting Week:	Estimated Number of Employees:
«MineIDNumber»	«ReportingWeek»	Between «EstimatedEmpMin» and «EstimatedEmpMax»
Mine Name:	Start With Number:	Take Every Number:
«MineName»	«StartWithNumber»	«TakeEveryNumber»

EMPLOYEE SELECTION INSTRUCTIONS

The Employee Questions ask you to report the demographic characteristics of a sample of your employees. This page contains instructions for selecting the sample of employees to include in the Employee Questions. (Please DO NOT include independent contractor employees in this part of the questionnaire, and DO NOT include any mine employee who was not at work during the REPORTING WEEK.)

Step 1. Print or copy a list from your files of the names and job titles of all mine employees who worked during the REPORTING WEEK (which includes the date shown in the box above) for the mining operation associated with the Mine ID and name (shown above). (Hourly and salaried employees can be combined, or listed separately, on the REPORTING WEEK list.)

Step 2. Sequentially number the salaried and hourly employees on your list, starting with the first name on the top of the list, e.g., 1, 2, 3, … This number will be the *employee sequence number*. [NOTE: The sequential numbering may be done by computer.]

Step 3. Record the total number of employees who worked during the REPORTING WEEK.

→ _____ = **TOTAL NUMBER WHO WORKED DURING REPORTING WEEK**

If this total number is . . .

... equal to 0, [not applicable] Go to Page 19.	... **fewer than 30**, *[all are to be included]* Please circle every one of the numbers you have recorded in Step 2, and **Go to next page**.	... **30 or more**, *[select a sample]* **Continue with Step 4**

Step 4. Quarterly reports indicate that this mine employs the Estimated Number of Employees shown in the box above. Does the number of employees recorded in Step 3 fall within the range of Estimated Number of Employees shown in the box above?

☐ Yes → **Continue with Step 5**.
☐ No → **If estimated number is incorrect, please call 1-888-814-4707 for assistance**. This is the toll-free number for Westat, the survey contractor.

Step 5. In these next steps, you will circle the *employee sequence numbers* for employees to be selected for the survey. To do this, you will use the **Start With Number** and **Take Every Number** printed in the box above.

Step 6. First, circle the *employee sequence number* that matches the **Start With Number** in the box above. This is the first employee selected for the survey.

Step 7. Next, start counting the *employee sequence numbers*, beginning with the sequence number after the one just circled. Count until you reach the **Take Every Number** listed in the box above. Circle that *employee sequence number*. This is the next selection.

Step 8. Repeat Step 7 until you come to the end of your employee list.

> **EXAMPLE: If total employees = 49, Start With Number = 2, and Take Every Number = 3, then you would circle the following employee sequence numbers:** 2, 5, 8, 11, 14, 17, 20, 23, 26, 29, 32, 35, 38, 41, 44, 47.

Step 9. Refer to the detailed instructions on the next page and record the sequence numbers you have circled in the first column of the Employee Questions.

INSTRUCTIONS FOR EMPLOYEE QUESTIONS

This section provides you with an item-by-item explanation for the Employee Questions. Please read these instructions carefully before completing the fold-out answer form on Page 17, or Employee Question screens on web version.

E1. Employee sequence number

This is the circled number from your employee roster list.

- If there are **fewer than 30** employees who worked during the REPORTING WEEK at your mine, all employees are included in the survey. Write each circled number on a separate line and provide the information corresponding to that employee.

- If there are **30 or more** employees who worked during the REPORTING WEEK at your mine, according to Steps 5-9 of the selection instructions, you have circled and recorded the sequence numbers of the employees being sampled. For example, if John Doe is fifth on your list, and he is selected to be included in the employee survey, then write "5" as the employee sequence number, and provide the information corresponding to that employee.

E2. Employee's regular job title

Regular job title means the title that specifies the employee's current position in the mine structure (e.g., manager). This information may be in an employee's personnel file or in the payroll system.

E3. Months or years of experience in this job title

Experience in this job title means the number of months or years that this employee has had his or her current job title. Report months only for those employees with less than 1 year of experience.

- **Months (MM) Column**: If the employee has been in the current job title **less than a year** at this mine, please **record** the number of **months** in the month's column. Round partial months up if one-half or more.

- **Years (YY) Column**: If the employee has been in the current job title **1 year or more**, please **record** number of **years** in the year's column. Round partial years up if one-half or more.

E4. Months or years of experience in this mine

Experience in this mine means the number of months or years that this employee has been working at this mine, from the time that the mine hired him or her. Report months only for those employees with less than 1 year of experience.

- **Months (MM) Column**: If the employee has worked for the mine **less than a year**, please **record** the number of **months** in the month's column. Round partial months up if one-half or more.

- **Years (YY) Column**: If the employee has worked for the mine **1 year or more**, please **record** number of **years** in the year's column. Round partial years up if one-half or more.

E5. Months or years of total mining experience

Total mining experience means the number of months or years that an employee has been employed in the mining industry overall. Please include years spent at other mining companies and at other ranks or job titles. Report months only for those employees with less than 1 year of experience.

- **Months (MM) Column**: If the employee has worked in the mining industry **less than a year**, please **record** the number of **months** in the month's column. Round partial months up if one-half or more.

- **Years (YY) Column**: If the employee has worked in the mining industry **1 year or more**, please **record** number of **years** in the year's column. Round partial years up if one-half or more.

INSTRUCTIONS FOR EMPLOYEE QUESTIONS (continued)

E6. Number of hours worked during the REPORTING WEEK

Number of hours worked means the number of hours for which the employee was paid conducting mining business during the REPORTING WEEK. The REPORTING WEEK includes the date shown in the box at the top of Page 1 or Page 14.

- **Do not include** vacation time, sick time, medical leave, or other time spent on non-work activities.

This information may be found in the employee's time reporting records.

E7. Employee's primary work location

Primary work location means the location where this employee worked the most hours in the REPORTING WEEK.

- Check ONLY one location.

Location categories (listed on the answer form/screen) are adapted from MSHA's Quarterly Mine Employment and Coal Production Report (MSHA Form 7000-2) with the exception that the following operational subunits have been combined into one work location: Auger, Culm Bank or Refuse Pile. This information may be found in the same employee work records that are used as source data to compile the MSHA quarterly report.

E8. Gender

Please specify by checking if the employee is male (M) or female (F). This information may be found in the employee's personnel file.

E9. Hispanic or Latino

Please specify ethnicity by checking whether or not the employee is Hispanic or Latino. Note that an additional question on the employee's race follows in the next question.

This information may be found in the employee's personnel file.

E10. Race

Please specify the employee's race by checking one or more categories.

E11. Year of birth

Please record the employee's year (YY) of birth. Use two digits for year (e.g., 1980 is "80").

This information may be found in the employee's personnel file.

E12. Highest level of education completed

Please check one category for *highest level of education completed*. This means the last grade that the employee completed.

This information is most likely included in the employee's personnel file.

Employee Questions

E1.	E2.	E3.		E4.		E5.		E6.	E7.									E8.		E9.		E10.					E11.	E12.				
Employee Sequence Number (from employee roster list)	Employee's Regular Job Title	Months **OR** Years of Experience						Number of Hours Worked during the REPORTING WEEK	EMPLOYEE'S PRIMARY WORK LOCATION Where He/She Worked the Most Hours in the REPORTING WEEK (Check ONLY One)									Gender		Hispanic or Latino		Race (Check One or More)					Year of Birth	Highest Level of Education Completed (Check ONLY One)				
		In this Job Title		In this Mine		Total Mining			**Underground Mine:** Underground	**Underground Mine:** Surface Shops, Yards, etc.	**Surface Mine** (including associated shops and yards): Strip, Open Pit, or Quarry	**Surface Mine** (including associated shops and yards): Auger, Culm Bank or Refuse Pile (Coal Mine Only)	**Surface Mine** (including associated shops and yards): Dredge	**Surface Mine** (including associated shops and yards): Other Surface Mining (Metal/ Nonmetal only)	Independent Shops or Yards	**Mill Operations, Preparation Plants, or Breakers** (include associated shops and yards)	**Office** (professional and clerical employees at the mine or plant working in an office)	M	F	Yes	No	American Indian or Alaska Native	Asian	Black or African American	Native Hawaiian or Other Pacific Islander	White	19YY	Less than 9th grade	9th-12th grade (no diploma)	HS Graduate or Equivalent (GED)	Some College, Associate Degree, or Technical School	Bachelor's Degree or beyond
		IF LESS THAN A YEAR, Enter Number of **Months**	IF ONE YEAR OR MORE, Enter Number of **Years**	IF LESS THAN A YEAR, Enter Number of **Months**	IF ONE YEAR OR MORE, Enter Number of **Years**	IF LESS THAN A YEAR, Enter Number of **Months**	IF ONE YEAR OR MORE, Enter Number of **Years**																									

Employee Questions

E1.		Employee Sequence Number (from employee roster list)	
E2.		Employee's Regular Job Title	
E3.	Months OR Years of Experience	In this Job Title	IF LESS THAN A YEAR, Enter Number of **Months**
			IF ONE YEAR OR MORE, Enter Number of **Years**
E4.		In this Mine	IF LESS THAN A YEAR, Enter Number of **Months**
			IF ONE YEAR OR MORE, Enter Number of **Years**
E5.		Total Mining	IF LESS THAN A YEAR, Enter Number of **Months**
			IF ONE YEAR OR MORE, Enter Number of **Years**
E6.		Number of Hours Worked during the REPORTING WEEK	
E7.	EMPLOYEE'S PRIMARY WORK LOCATION Where He/She Worked the Most Hours in the REPORTING WEEK (Check ONLY One)		**Underground Mine:** Underground
			Underground Mine: Surface Shops, Yards, etc.
			Surface Mine (including associated shops and yards): Strip, Open Pit, or Quarry
			Surface Mine (including associated shops and yards): Auger, Culm Bank or Refuse Pile (Coal Mine Only)
			Surface Mine (including associated shops and yards): Dredge
			Surface Mine (including associated shops and yards): Other Surface Mining (Metal/Nonmetal only)
			Independent Shops or Yards
			Mill Operations, Preparation Plants, or Breakers (include associated shops and yards)
			Office (professional and clerical employees at the mine or plant working in an office)
E8.		Gender	M
			F
E9.		Hispanic or Latino	Yes
			No
E10.		Race (Check One or More)	American Indian or Alaska Native
			Asian
			Black or African American
			Native Hawaiian or Other Pacific Islander
			White
E11.		Year of Birth	19YY
E12.		Highest Level of Education Completed (Check ONLY One)	Less than 9th grade
			9th-12th grade (no diploma)
			HS Graduate or Equivalent (GED)
			Some College, Associate Degree, or Technical School
			Bachelor's Degree or beyond

FINAL QUESTIONS AND COMMENTS

F1. In the REPORTING WEEK, were there any events or circumstances that would make what you have reported unusual (e.g., severe weather conditions, trouble in production, a labor strike, etc.)?

☐ Yes → Go to Question F2
☐ No → Go to Question F3

F2. [IF YES TO Question F1]: Please specify the unusual events:

_____

_____

_____

F3. What is today's date? |__|__| |__|__| |__|__|__|__|
                                              M M   D D   Y Y Y Y

F4. Please make a copy of this completed questionnaire and your list of sampled employees (keep these on file for 60 days) in case we need to contact you for clarification.

F5. Please provide the company representative to be contacted regarding the completion of the questionnaire:

    Name: _____

    Title: _____

    Telephone: (   ) _____

F6. Reminder: If you so indicated in question M22, please enclose an example of your mine schedule with your completed questionnaire.

F7. Please mail this completed questionnaire in the provided business reply envelope to the survey contractor: **Westat, Room TC-1046F, 1650 Research Boulevard, Rockville, MD 20850-3195.**

Please record any comments on the next page.

FINAL QUESTIONS AND COMMENTS (continued)

F8. Do you have any comments, or is there any other information you can provide that may help us understand the answers you provided? (Please include question numbers for comments or explanations related to specific responses.)

___

Thank you for your participation in this survey!

Delivering on the Nation's promise:
Safety and health at work for all people
through research and prevention

*If you have any questions regarding the
National Survey of the Mining Population, please contact:*

Linda J. McWilliams
Project Director
NIOSH, Pittsburgh Research Laboratory
P.O. Box 18070
626 Cochrans Mill Road
Building 01
Pittsburgh, PA 15236

Telephone: (412) 386-6116
Fax: (412) 386-6780
E-mail: LMcWilliams@cdc.gov

http://www.cdc.gov/niosh/mining/statistics/survey.htm

Appendix B. Questions and Answers Brochure

National Survey of the Mining Population

Questions and Answers

Sponsored by the
National Institute for Occupational
Safety and Health (NIOSH)
Pittsburgh Research Laboratory
P.O. Box 18070
Pittsburgh, PA 15236

For further information on the purpose of this survey, please contact:

Linda McWilliams
Project Director
NIOSH, Pittsburgh Research Laboratory
P.O. Box 18070
626 Cochrans Mill Road
Building 01
Pittsburgh, PA 15236
(412) 386-6116
E-mail: LMcWilliams@cdc.gov

http://www.cdc.gov/niosh/mining/statistics/survey.htm

For further information on how to fill out the questionnaire, please contact:

Westat
Attn: National Mining Survey
1650 Research Boulevard
Room TC-1046F
Rockville, MD 20850
(888) 814-4707

Do I need to report data for all employees of our mining operation?

If you have **less than 30** employees at this mining operation, we ask you to report for all of them.

If you have **30 or more** employees at this mine, we ask you to report data for only a sample of them.

In order to get good data about the mining industry, it is very important that you sample accurately. Our aim is to make the sampling of employees as simple as possible. Step by step instructions are provided in the survey booklet.

Should I include independent contractors in the employee questions?

Contractor information should only be included when responding to the **mine questions**. Contractors should *not* be counted as employees when completing the **employee questions**. Only data for mine operator employees should be included on the employee questions.

If the mining operation is being run by your company under contract to the owner, report for your employees but exclude workers associated with other independent contractors.

How long will the survey take?

Although this varies by mining operation, on average it will take 120 minutes to complete the survey. This includes obtaining information from personnel records, and should take less time for smaller mines.

Why is this survey being done?

The mission of the National Institute for Occupational Safety and Health (NIOSH) is working to improve the safety and health of American workers. As part of this effort, NIOSH/Pittsburgh Research Laboratory (PRL) is collecting demographic and other data on the mining industry.

Since 1986, there has been little research on the demographics of the mining labor force, such as age, gender, job title, languages used, educational attainment, race, ethnicity, and years of mining experience. These data are needed to understand the risk of work-related injuries, disease, and fatalities and to customize safety and health interventions for specific groups of the mining industry. These data can also be used to learn more about the underlying causes of work-related incidents and to identify ways to reduce their occurrence.

NIOSH/PRL is sponsoring this survey of mining operations and their employees to fill this data gap. Our main objectives are to:

▲ collect basic information about mining operations;

▲ establish the demographic and occupational characteristics of mine operator employees for each mining commodity (i.e., coal, metal, nonmetal, stone, and sand and gravel); and

▲ estimate the number of independent contractor employees used by mining operations and their occupational characteristics.

What will the mining industry and my mine get out of this survey?

The ultimate goal of the survey is to minimize and prevent work-related injuries and diseases that harm miners and reduce productivity. NIOSH will use the information you provide to clarify safety and health issues and calculate injury rates for various occupations. For example, we now know how many electricians are reported as injured in mine accidents, but we don't know how many total electricians work in the mining industry, in order to calculate their injury rates. Once the survey is completed, such rates will be available, and NIOSH will send you a copy of the final report.

What data will be collected?

There are two sets of data being collected:

▲ The **mine questions** include items about the mining operation, its use of independent contractors, safety, and communication measures.

▲ The **employee questions** include demographic and occupational questions about individual mine employees.

It is important that you complete *both* parts of the survey. You have the option of completing either the survey questionnaire booklet or an Internet survey questionnaire. Both versions ask the same questions. Instructions to access the Internet questionnaire are attached to the cover letter included in this mailing.

Am I required to participate?

Your participation is voluntary and you may refuse to answer any question for any reason. However, the participation of each selected mining operation is vital to the success of the survey.

Why was my mining operation chosen?

Your mining operation was randomly selected from a list of all mining operations nationwide. The sample must represent the diversity of mining operations across the Nation. The information you provide is essential to obtain an accurate picture of the mining industry.

Who will see my responses?

Only NIOSH researchers, and researchers from Westat, NIOSH's data collection contractor, will see your responses. Both organizations are firmly committed to protecting the survey data and will not release this information unless compelled by law. The answers from all participating mines will be published only as summarized data so that no single company or individual employee can be identified.

Is it appropriate for me to release information about employees who work here?

You will not be reporting the names or other identifying information of individual employees. The data you provide cannot be linked to any of your individual employees.

Appendix C. MSHA Form 7000-2: Quarterly Mine Employment and Coal Production Report

Quarterly Mine Employment
and Coal Production Report
(SEE INSTRUCTIONS ON REVERSE SIDE OF COPY 2)

DOL - MSHA - PEIR - OIEI
P.O. Box 25367
Denver, Colorado 80225 - 0367

Date Report Completed

Mo. _____ Day _____ Yr. _____

Mail Before

1. Persons Working, Employee-Hours, and Coal Production

(1) Operation Sub Unit Code(s) previously reported:	Code	(2) Average number of persons working during quarter	(3) Total employee hours worked during the quarter	(4) Production of clean coal during quarter, (short tons)
Underground Mine — Underground	01			
Surface Shops, Yards, etc.	02			
Surface Mine — Strip, Open Pit, or Quarry	03			
(including associated shops and yards) — Auger (Coal Mine Only)	04			
Culm Bank or Refuse Pile (Coal Mine Only)	05			
Dredge	06			
Other Surface Mining (Metal/Nonmetal Only)	12			
Independent Shops or Yards	17			
Mill Operations, Preparation Plants, or Breakers (include associated shops and yards)	30			
Office (professional and clerical employees at the mine or plant working in an office)	99			

2. Other Reportable Data

How many MSHA reportable injuries or illnesses did you have this quarter? ____

Person to be contacted regarding this report:

Name _____

Title _____

Tel. No. (___) ___ - ____
            area code

For _____ Quarter _____ Year _____

Check here if this report is being submitted by a contractor ☐

If any information below is incorrect, please enter correct information here:

County: _____

Operation Name: _____

Operating Company Name and Mailing Address: _____

County _____

MSHA ID Number _____ Contractor ID _____

Operation Name _____

Operating Company Name and Mailing Address _____

MSHA Form 7000-2, July 97, (revised)

OMB Number 1219-0007; Approval Expires April 30, 2011

Copy 1 - Return to MSHA (Denver)

Quarterly Mine Employment and Coal Production Report

U.S. Department of Labor

Mine Safety and Health Administration

OBM Control Number 1219-0007; Approval Expires April 30, 2011

This report is required by law (30 U.S.C. subsection 813; 30 C.F.R. Part 50). Failure to report may result in the issuance of a citation or order under 30 U.S.C. subsection 814 to an operator of a coal or other mine, the assessment of a civil penalty against an operator of a coal or other mine under 30 U.S.C. subsection 820(a), and the institution of a civil action under 30 U.S.C. subsection 818. An individual who knowingly makes a false statement in any report shall, upon conviction, be punished by a fine of not more than $10,000 or by imprisonment for not more than 5 years, or both, under 30 U.S.C. subsection 820(f). Whoever, in any matter within the jurisdiction of any department or agency of the United States knowingly and willfully falsifies, conceals or covers up by any trick scheme, or device, a material fact, or makes or uses any false writing or document knowing the same to contain any false, fictitious or fraudulent statement or entry, shall be fined under 18 U. S. C. or imprisoned not more than five years, or both, under 18 U. S. C. subsection 1001.

Important: (INSTRUCTIONS)

This form must be completed and mailed or faxed within 15 days after the end of each calendar quarter.

1. Fill out this form as completely as possible and return Copy 1 of this report to:
   MSHA
   PEIR - Office of Injury and Employment Information   **OR**   You may FAX Copy 1 to Fax # 1- 888 - 231 - 5515
   P.O. BOX 25367
   Denver, CO 80225-0367
2. If it is necessary to make any address changes, indicate correct information on this form.
3. When pre-addressed, this form is only for the operation with I. D. number as shown. Do not use for any other operation.
4. **Sand and Gravel** operators report employment data under code 03 or 06 as appropriate, except for data on office workers which should be reported under code 99.
5. All mine operators and independent contractors reporting as required by 30 C.F.R. Part 50, should show persons working and employee hours worked; those producing coal should also show production date.
6. **Independent Contractors** should complete quarterly only **one** form for activities at all coal locations, and one form for activities at metal and nonmetal locations.

The public reporting burden for this collection of Information is estimated to average 30 minutes per response, including the time for reviewing instructions, searching existing date sources, gathering and maintaining the date needed, and completing and reviewing the collection of Information. Send comments regarding this estimated response time or any other aspect of this collection of information, including suggestions for reducing this burden, to Mine Safety and Health Administration, U.S. Department of Labor, 1100 Wilson Boulevard, Arlington, VA 22209-3939.

Persons are not required to respond to this collection of information unless this form displays a currently valid OMB control number.

MSHA Form 7000-2, July 97 (revised)

Appendix D. Standard Industrial Classification (SIC) for Active Mines in 2007

Coal Mining Sector
- Anthracite Coal
- Bituminous Coal

Metal Mining Sector
- Alumina (Mill)
- Aluminum Ore
- Beryl
- Chromite
- Copper Ore
- Gold (Lode & Placer)
- Iron Ore
- Lead and/or Zinc Ore
- Manganese
- Metal Ores, NEC
- Molybdenum
- Platinum Group
- Rare Earths
- Silver Ores
- Titanium
- Uranium
- Uranium – Vanadium Ores
- Vanadium
- Zircon

Nonmetal Mining Sector
- Aplite
- Barite
- Boron Materials
- Brucite
- Chemical and Fertilizer, NEC
- Clay (Common)
- Clay (Fire)
- Clay, Ceramic and Refractory, NEC
- Feldspar
- Fluorspar
- Gemstones
- Gilsonite
- Gypsum
- Kyanite

Nonmetal Mining Sector (Cont.)
- Leonardite
- Magnesite
- Mica
- Nonmetallic Minerals, NEC
- Oil Sand
- Oil Shale
- Perlite
- Phosphate Rock
- Pigment Mineral
- Potash
- Potash, Soda, & Borate Minerals, NEC
- Pumice
- Salt (Evaporated)
- Salt (Rock)
- Shale (Common)
- Sodium Compounds
- Talc, Soapstone, & Pyrophyllite
- Trona
- Vermiculite

Stone Mining Sector
- Cement
- Granite (Crushed & Broken)
- Granite (Dimension)
- Lime
- Limestone (Crushed & Broken)
- Limestone (Dimension)
- Marble (Crushed & Broken)
- Marble (Dimension)
- Sandstone (Crushed & Broken)
- Sandstone (Dimension)
- Slate (Crushed & Broken)
- Slate (Dimension)
- Stone, Crushed & Broken, NEC
- Stone, Dimension, NEC
- Traprock (Crushed & Broken)
- Traprock (Dimension)

Sand and Gravel Mining Sector
- Sand & Gravel

Abbreviation: NEC, not elsewhere classified

Appendix E. Stratification and Sample Size Guidelines

Stratification

The cum $\sqrt{f}$ rule is often suggested for use in forming strata for surveys of businesses, which typically have a large number of small businesses with very few employees and a small number of large businesses with quite substantial payrolls [Cochran 1977]. Using this approach, strata that have approximately equal sizes in terms of the square root of the size measure are established. The cum $\sqrt{f}$ rule was used in determining the initial size-based strata for each mining sector with an assumption of about 4–5 strata per sector for underground mines and for surface mines. Except for sand and gravel mines, the large mines account for 25 percent or more of total employment. These initial stratum definitions for each commodity varied somewhat across mining sectors but were similar.

The next step in stratum formation was to recognize that data from the five mining sectors would be combined to study mining as a whole. Using common definitions for strata across the five sectors facilitated these combined analyses. The initial stratum definitions were compared to determine a common stratification approach. The stratum definitions that met the needs for all five commodities were formed by the cross of underground versus surface mines with these size groupings of employees: 1–9, 10–25, 26–50, 51–75, 76–100, 101–250, and 251 and up.

Sample Size

To determine the stratum sample sizes, the precision of percentage estimates under various sample sizes was considered. Table E-1 presents the half-length of confidence intervals around an estimated percentage $\hat{P}$ under various sample size and design effects and assuming large population sizes. For this table, the confidence interval was approximated for design purposes as:

$$\hat{P} \pm z_{1-\alpha} \sqrt{Var(\hat{P})} \qquad (1)$$

Here $z_{1-\alpha}$ is the value of the critical point x at which the normal cumulative distribution function equals 1-α, and $Var(\hat{P})$ is the variance of $\hat{P}$. The half-length HL is:

$$HL = z_{1-\alpha} \sqrt{Var(\hat{P})} \qquad (2)$$

That is, $\hat{P}$ can be expected to fall within the range [$P-HL$, $P+HL$] with 95 percent confidence for the indicated sample sizes.

To determine these half-lengths of confidence intervals, there is a need to estimate the variance of the estimated percentage $\hat{P}$. Ignoring finite population correction factors, Table E-1 models the variance for an estimated percentage $\hat{P}$ as:

$$Var(\hat{P}) = \frac{P(100-P)}{n} DEFF \qquad (3)$$

where n is the sample size, P is the percentage being estimated, and $DEFF$ is the design effect. The design effect for a survey estimate is defined to be the ratio of the statistic under the actual design divided by the variance that would have been achieved from a simple random sample of the same size. The design effect represents the cumulative effect of design components such as stratification, unequal weighting, and clustering, and varies with each design. The design effects for this survey were estimated to be about 1.00 for mine-level and employee-level estimates within strata. Crosscutting estimates were likely to have larger design effects, particularly for employee-level estimates. The design effect differs from 1.00 for the crosscutting estimates due to the variation in sampling rates used across strata. Fortunately, these crosscutting estimates often have large sampler sizes due to combining samples across strata.

Sample sizes were set with the guideline that the precision for stratum estimates was constrained as that shown for sample sizes of 100 in Table E-1. Some mine strata have very small population sizes and some mining sectors are small overall. In such situations, the variance as given in equation (3) is reduced by the factor $(N-n)/(N-1)$, where n is the sample size and N is the population size. Rather than create versions of Table E-1 for all possible population sizes, finite-population-corrected (fpc) sample sizes were developed. An actual sample size of n for a population of size N is equivalent to the precision achieved with a sample size of $n' = \frac{n(N-1)}{N-n}$ from a population so large that fpc effects are ignorable. Initial sample sizes were set for each stratum so that the finite-population-corrected sample size was about 100 and then inflated to account for a projected 80 percent response rate. These initial sample sizes were then adjusted to prevent excessive variations in the sampling rates across strata for mines and for employees.

Besides the number of mines selected, the employee sample size is affected by the eligibility and response rates for mines and the average number of employees sampled per mine. The average number of employees sampled per mine would be about 20 except for the smallest stratum where approximately 5 employees would tend to be sampled. It was assumed that 80 percent of all eligible mines would respond, providing both mine-level and employee-level data. For sample design purposes, the assumption was made that a variable percentage of mines would be eligible for the survey, depending upon employment size. An eligibility rate of 85 percent was assumed for mines with 1–9 employees. These mines are most likely to shut down operations or go out of business. An eligibility rate of 90 percent was assumed for mines with 10–50 employees, and 95 percent for mines with 51–100 employees. For very large mines with employment of more than 100, an eligibility rate of 99 percent was assumed, as they should be most stable in terms of their operations.

In designing the commodity samples, an effort was made to minimize the design effects for mine-level and employee-level analyses. In particular, the goal was to achieve design effects of 1.0 for within-stratum estimates and design effects of 2.0 or less for crosscutting estimates. Following standard practice, the design effect $DEFF$ was modeled as the product of the design effect associated with unequal weighting D_w and the design effect for clustering D_c, that is

$DEFF = D_w * D_c$. A simple random sample has both design effect components equal to one—therefore $DEFF=1$.

Both mine-level and employee-level estimates could potentially be subject to an unequal weighting effect greater than one, particularly for crosscutting estimates that combine data from multiple strata. The design effect for unequal weighting can be estimated as:

$$D_w = \frac{n \sum_{i=1}^{n} W_i^2}{\left(\sum_{i=1}^{n} W_i\right)^2} \quad (4)$$

where n is the total sample size and W_i is the weight for the ith observation. When the weights (the inverse of the selection probabilities) are equal for all selections, $D_w = 1$. For mines, $D_w = 1$ within all strata for the proposed designs and was often only slightly greater than one across strata. For employees, $D_w = 1$ except for the two largest strata that collapsed employee size categories. These strata tended to have all mines selected with certainty, so the only way to reduce D_w was to increase the number of employees sampled per mine from 25 to 50. Adjusting the sample size for the very large mines could even out the employee-level weights within these strata and across strata. However, the increase in employee sample size also increased the burden for the mine respondent and increased the design effect for clustering.

The design effect associated with clustering measures the loss of precision of a clustered sample as compared with a simple random sample. Clustered samples tend to have less precision than simple random samples of the same size, because units within the same cluster usually are more homogeneous than units from different clusters. The design effect for clustering can be estimated as:

$$D_c = 1 + \rho(b-1) \quad (5)$$

where ρ is the intracluster correlation coefficient and b is the cluster size. Because stratified simple random sampling would be used to select mines, the mines would not be clustered ($b = 1$) and mine-level estimates would not be subject to a clustering effect ($D_c = 1$). However, multiple employees would be selected from each mine, so employee-level estimates would be subject to a design effect due to clustering. For the purpose of modeling the clustering design effect, it was assumed that variable values for ρ be based upon the size of the mine. Employees within small mines with 1 to 50 employees were expected to be more homogeneous, so a value of $\rho = 5$ percent was assumed. Medium size mines were assumed to be less homogeneous, so a value of $\rho = 3$ percent was assumed. Large mines with more than 100 employees were expected to be quite diverse, so a value of $\rho = 1$ percent was assumed. A value of $\rho = 3$ percent was assumed for estimates compiled across strata.

Table E-1. Half-Length of 95% Confidence Intervals in Percentage Points for Various Percentages Being Estimated for Domains of Various Sizes with Various Design Effects

DEFF	P	50	75	100	150	200	250	350	400	500
1.00	10	8	7	6	5	4	4	3	3	3
1.00	20	11	9	8	6	6	5	4	4	4
1.00	25	12	10	8	7	6	5	5	4	4
1.00	30	13	10	9	7	6	6	5	4	4
1.00	40	14	11	10	8	7	6	5	5	4
1.00	50	14	11	10	8	7	6	5	5	4
1.25	10	9	8	7	5	5	4	4	3	3
1.25	20	12	10	9	7	6	6	5	4	4
1.25	25	13	11	9	8	7	6	5	5	4
1.25	30	14	12	10	8	7	6	5	5	4
1.25	40	15	12	11	9	8	7	6	5	5
1.25	50	15	13	11	9	8	7	6	5	5
1.50	10	10	8	7	6	5	5	4	4	3
1.50	20	14	11	10	8	7	6	5	5	4
1.50	25	15	12	10	8	7	7	6	5	5
1.50	30	16	13	11	9	8	7	6	6	5
1.50	40	17	14	12	10	8	7	6	6	5
1.50	50	17	14	12	10	8	8	6	6	5
2.00	10	12	10	8	7	6	5	4	4	4
2.00	20	16	13	11	9	8	7	6	6	5
2.00	25	17	14	12	10	8	8	6	6	5
2.00	30	18	15	13	10	9	8	7	6	6
2.00	40	19	16	14	11	10	9	7	7	6
2.00	50	20	16	14	11	10	9	7	7	6
3.00	10	12	10	8	7	6	5	4	4	4
3.00	20	16	13	11	9	8	7	6	6	5
3.00	25	17	14	12	10	8	8	6	6	5
3.00	30	18	15	13	10	9	8	7	6	6
3.00	40	19	16	14	11	10	9	7	7	6
3.00	50	20	16	14	11	10	9	7	7	6

Appendix F. Sample Size Allocation Using MSHA Data from the Second Quarter of 2002

Table F-1. Sample Allocation for Underground Coal Mines

Stratum	Number of Mines	Percentage of Total Mines	Number of Employees	Percentage of Total Employees	Sample Mines	Eligibility Rate	Response Rate	Responding Eligible Mines
1–9	102	19%	461	1%	56	85%	80%	38
10–25	149	27%	2,589	7%	68	90%	80%	49
26–50	146	26%	5,206	15%	67	90%	80%	48
51–75	49	9%	3,098	9%	35	95%	80%	27
76–100	22	4%	1,917	5%	22	95%	80%	17
101–250	49	9%	8,301	24%	49	99%	80%	39
251+	34	6%	13,477	38%	34	99%	80%	27
Total	551	100%	35,049	100%	331			244

Stratum	Employees Sampled Per Mine	Total Sample Employees	Nonresponse Adjusted Mine Weight	Average Employee Weight	Mine DEFF	Employee D_w	Employee Eligibility Rate	Employee ρ	Employee D_c	Employee DEFF
1–9	5	172	2.3	2.3	1.0	1.0		5%	1.2	1.2
10–25	17	851	2.7	2.7	1.0	1.0		5%	1.8	1.8
26–50	18	860	2.7	5.4	1.0	1.0		5%	1.8	1.8
51–75	21	561	1.8	5.3	1.0	1.0		3%	1.6	1.6
76–100	22	364	1.3	5.0	1.0	1.0		3%	1.6	1.6
101–250	23	908	1.3	9.1	1.0	1.0		1%	1.2	1.3
251+	24	651	1.3	20.5	1.0	1.1		1%	1.2	1.3
Total		4,366			1.1	1.6		3%	1.5	2.5

Table F-2. Sample Allocation for Surface Coal Mines

Stratum	Number of Mines	Percentage of Total Mines	Number of Employees	Percentage of Total Employees	Sample Mines	Eligibility Rate	Response Rate	Responding Eligible Mines
1–9	518	46%	2,193	6%	101	85%	80%	69
10–25	252	23%	4,166	12%	84	90%	80%	60
26–50	188	17%	6,860	19%	75	90%	80%	54
51–75	58	5%	3,500	10%	36	95%	80%	27
76–100	24	2%	2,068	6%	20	95%	80%	15
101–250	52	5%	8,114	23%	52	99%	80%	41
251+	23	2%	8,823	25%	23	99%	80%	18
Total	1,115	100%	35,724	100%	391			285

Stratum	Employees Sampled Per Mine	Total Sample Employees	Nonresponse Adjusted Mine Weight	Average Employee Weight	Mine DEFF	Employee D_w	Employee ρ	Employee D_c	Employee DEFF
1–9	4	291	6.4	6.4	1.0	1.0	5%	1.2	1.2
10–25	17	1,000	3.8	3.8	1.0	1.0	5%	1.8	1.8
26–50	18	985	3.1	6.3	1.0	1.0	5%	1.9	1.9
51–75	20	550	2.0	6.0	1.0	1.0	3%	1.6	1.6
76–100	22	327	1.5	6.0	1.0	1.0	3%	1.6	1.6
101–250	23	935	1.3	8.6	1.0	1.1	1%	1.2	1.3
251+	25	461	1.3	19.0	1.0	1.1	1%	1.2	1.4
Total		4,549			1.3	1.4	5%	1.7	2.4

Table F-3. Sample Allocation for Underground Metal Mines

Stratum	Number of Mines	Percentage of Total Mines	Number of Employees	Percentage of Total Employees	Sample Mines	Eligibility Rate	Response Rate	Responding Eligible Mines
1–9	18	35%	92	2%	18	85%	80%	12
10–25	7	14%	123	2%	7	90%	80%	5
26–50	4	8%	171	3%	4	90%	80%	3
51–75	2	4%	125	2%	2	95%	80%	2
76–100	3	6%	264	5%	3	95%	80%	2
101–250	12	24%	1,844	36%	12	99%	80%	10
251+	5	10%	2,476	49%	5	99%	80%	4
Total	**51**	**100%**	**5,095**	**100%**	**51**			**37**

Stratum	Employees Sampled Per Mine	Total Sample Employees	Nonresponse Adjusted Mine Weight	Average Employee Weight	Mine DEFF	Employee D_w	Employee ρ	Employee D_c	Employee DEFF
1–9	5	63	1.3	1.3	1.0	1.0	5%	1.2	1.2
10–25	18	89	1.3	1.3	1.0	1.0	5%	1.8	1.8
26–50	21	62	1.3	2.5	1.0	1.0	5%	2.0	2.0
51–75	21	32	1.3	3.8	1.0	1.0	3%	1.6	1.6
76–100	22	50	1.3	5.0	1.0	1.0	3%	1.6	1.6
101–250	23	222	1.3	8.2	1.0	1.1	1%	1.2	1.3
251+	24	96	1.3	25.4	1.0	1.4	1%	1.2	1.7
Total		**613**			**1.0**	**2.5**	**3%**	**1.5**	**3.7**

Table F-4. Sample Allocation for Surface Metal Mines

Stratum	Number of Mines	Percentage of Total Mines	Number of Employees	Percentage of Total Employees	Sample Mines	Eligibility Rate	Response Rate	Responding Eligible Mines
1–9	54	34%	235	1%	54	85%	80%	37
10–25	27	17%	438	2%	27	90%	80%	19
26–50	10	6%	356	2%	10	90%	80%	7
51–75	9	6%	591	3%	9	95%	80%	7
76–100	12	7%	1,094	5%	12	95%	80%	9
101–250	19	12%	2,959	13%	19	99%	80%	15
251+	30	19%	17,703	76%	30	99%	80%	24
Total	161	100%	23,376	100%	161			118

Stratum	Employees Sampled Per Mine	Total Sample Employees	Nonresponse Adjusted Mine Weight	Average Employee Weight	Mine DEFF	Employee D_w	Employee ρ	Employee D_c	Employee DEFF
1–9	4	160	1.3	1.3	1.0	1.0	5%	1.2	1.2
10–25	16	315	1.3	1.3	1.0	1.0	5%	1.8	1.8
26–50	18	128	1.3	2.5	1.0	1.0	5%	1.8	1.8
51–75	22	150	1.3	3.8	1.0	1.0	3%	1.6	1.6
76–100	23	208	1.3	5.0	1.0	1.0	3%	1.7	1.7
101–250	23	344	1.3	8.5	1.0	1.1	1%	1.2	1.3
251+	24	581	1.3	30.2	1.0	1.4	1%	1.2	1.7
Total		1,886			1.0	2.7	3%	1.4	3.9

Table F-5. Sample Allocation for Underground Nonmetal Mines

Stratum	Number of Mines	Percentage of Total Mines	Number of Employees	Percentage of Total Employees	Sample Mines	Eligibility Rate	Response Rate	Responding Eligible Mines
1–9	12	29%	50	1%	12	85%	80%	8
10–25	2	5%	38	1%	2	90%	80%	1
26–50	8	20%	290	6%	8	90%	80%	6
51–75	4	10%	232	5%	4	95%	80%	3
76–100	1	2%	80	2%	1	95%	80%	1
101–250	9	22%	1,634	34%	9	99%	80%	7
251+	5	12%	2,431	51%	5	99%	80%	4
Total	41	100%	4,755	100%	41			30

Stratum	Employees Sampled Per Mine	Total Sample Employees	Nonresponse Adjusted Mine Weight	Average Employee Weight	Mine DEFF	Employee D_w	Employee ρ	Employee D_c	Employee DEFF
1–9	4	34	1.25	1.3	1.0	1.00	5%	1.2	1.2
10–25	19	27	1.25	1.3	1.0	1.00	5%	1.9	1.9
26–50	18	104	1.25	2.5	1.0	1.00	5%	1.9	1.9
51–75	19	59	1.25	3.8	1.0	1.00	3%	1.6	1.6
76–100	20	15	1.25	5.0	1.0	1.00	3%	1.6	1.6
101–250	24	169	1.25	9.6	1.0	1.03	1%	1.2	1.3
251+	24	96	1.25	25.1	1.0	1.26	1%	1.2	1.5
Total		504			1.0	2.18	3%	1.1	2.4

Table F-6. Sample Allocation for Surface Nonmetal Mines

Stratum	Number of Mines	Percentage of Total Mines	Number of Employees	Percentage of Total Employees	Sample Mines	Eligibility Rate	Response Rate	Responding Eligible Mines
1–9	347	53%	1,454	8%	92	85%	80%	63
10–25	136	21%	2,094	12%	65	90%	80%	47
26–50	73	11%	2,768	15%	46	90%	80%	33
51–75	45	7%	2,799	16%	34	95%	80%	26
76–100	14	2%	1,191	7%	14	99%	80%	11
101–250	25	4%	3,790	21%	25	99%	80%	20
251+	10	2%	3,785	21%	10	99%	80%	8
Total	650	100%	17,881	100%	286			207

Stratum	Employees Sampled Per Mine	Total Sample Employees	Nonresponse Adjusted Mine Weight	Average Employee Weight	Mine DEFF	Employee D_w	Employee ρ	Employee D_c	Employee DEFF
1–9	4	262	4.71	4.7	1.0	1.00	5%	1.2	1.2
10–25	15	721	2.62	2.6	1.0	1.00	5%	1.7	1.7
26–50	19	628	1.98	4.0	1.0	1.00	5%	1.9	1.9
51–75	21	536	1.65	5.0	1.0	1.00	3%	1.6	1.6
76–100	21	236	1.25	5.0	1.0	1.00	3%	1.6	1.6
101–250	23	450	1.25	8.3	1.0	1.04	1%	1.2	1.3
251+	24	191	1.25	19.6	1.0	1.15	1%	1.2	1.4
Total		3,023			1.2	1.65	3%	1.0	1.6

Table F-7. Sample Allocation for Underground Stone Mines

Stratum	Number of Mines	Percentage of Total Mines	Number of Employees	Percentage of Total Employees	Sample Mines	Eligibility Rate	Response Rate	Responding Eligible Mines
1–9	20	18%	102	3%	20	85%	80%	14
10–25	48	43%	766	22%	35	90%	80%	25
26–50	28	25%	955	27%	23	90%	80%	17
51–75	10	9%	610	17%	10	95%	80%	8
76–100	1	1%	91	3%	1	95%	80%	1
101–250	3	3%	377	11%	3	99%	80%	2
251+	1	1%	637	18%	1	99%	80%	1
Total	111	100%	3,538	100%	93			67

Stratum	Employees Sampled Per Mine	Total Sample Employees	Nonresponse Adjusted Mine Weight	Average Employee Weight	Mine DEFF	Employee D_w	Employee ρ	Employee D_c	Employee DEFF
1–9	5	69	1.3	1.3	1.0	1.000	5%	1.2	1.2
10–25	16	402	1.7	1.7	1.0	1.000	5%	1.7	1.7
26–50	17	282	1.5	3.0	1.0	1.000	5%	1.8	1.8
51–75	20	155	1.3	3.8	1.0	1.000	3%	1.6	1.6
76–100	23	17	1.3	5.0	1.0	1.000	3%	1.7	1.7
101–250	22	53	1.3	7.1	1.0	1.007	1%	1.2	1.2
251+	25	19	1.3	32.5	1.0	1.000	1%	1.2	1.2
Total		998			1.0	2.710	3%	1.0	2.6

Table F-8. Sample Allocation for Surface Stone Mines

Stratum	Number of Mines	Percentage of Total Mines	Number of Employees	Percentage of Total Employees	Sample Mines	Eligibility Rate	Response Rate	Responding Eligible Mines
1–9	1,698	46%	8,067	11%	116	85%	80%	79
10–25	1,304	35%	20,497	28%	114	90%	80%	82
26–50	402	11%	13,862	19%	95	90%	80%	68
51–75	104	3%	6,356	9%	51	95%	80%	39
76–100	54	1%	4,704	6%	35	95%	80%	27
101–250	124	3%	17,528	24%	62	99%	80%	49
251+	6	0%	1,796	2%	6	99%	80%	5
Total	3,692	100%	72,810	100%	479			349

Stratum	Employees Sampled Per Mine	Total Sample Employees	Nonresponse Adjusted Mine Weight	Average Employee Weight	Mine DEFF	Employee D_w	Employee ρ	Employee D_c	Employee DEFF
1–9	5	375	18.3	18.3	1.0	1.000	5%	1.2	1.2
10–25	16	1,290	14.3	14.3	1.0	1.000	5%	1.7	1.7
26–50	17	1,179	5.3	10.6	1.0	1.000	5%	1.8	1.8
51–75	20	790	2.5	7.6	1.0	1.000	3%	1.6	1.6
76–100	22	579	1.9	7.7	1.0	1.000	3%	1.6	1.6
101–250	23	1,126	2.5	15.4	1.0	1.046	1%	1.2	1.3
251+	24	114	1.3	15.7	1.0	1.028	1%	1.2	1.3
Total		5,453			1.5	1.092	3%	1.0	1.1

Table F-9. Sample Allocation for Sand and Gravel Mines

Stratum	Number of Mines	Percentage of Total Mines	Number of Employees	Percentage of Total Employees	Sample Mines	Eligibility Rate	Response Rate	Responding Eligible Mines
1–3	2,589	42.6%	5,504	13.3%	119	85%	80%	81
4–6	1,572	25.9%	7,570	18.4%	80	85%	80%	54
7–9	748	12.3%	5,872	14.2%	37	85%	80%	25
10–25	963	15.9%	13,995	33.9%	110	90%	80%	79
26–50	168	2.8%	5,743	13.9%	70	95%	80%	53
51–75	27	0.4%	1,607	3.9%	16	95%	80%	12
76–100	3	0.0%	264	0.6%	3	99%	80%	2
101–250	4	0.1%	683	1.7%	4	99%	80%	3
251+	0	0.0%	0	0.0%	--	--	--	--
Total	6,074	100.0%	41,238	100%	439			311

Stratum	Employees Sampled Per Mine	Total Sample Employees	Nonresponse Adjusted Mine Weight	Average Employee Weight	Mine DEFF	Employee D_w	Employee ρ	Employee D_c	Employee DEFF
1–3	2	172	27	27	1.00	1.00	5%	1.1	1.06
4–6	5	262	25	25	1.00	1.00	5%	1.2	1.19
7–9	8	198	25	25	1.00	1.00	5%	1.3	1.34
10–25	15	1,151	11	11	1.00	1.00	5%	1.7	1.68
26–50	17	909	3	6	1.00	1.00	5%	1.8	1.80
51–75	20	241	2	6	1.00	1.00	3%	1.6	1.57
76–100	22	52	1	5	1.00	1.00	3%	1.6	1.63
101–250	24	75	1	9	1.00	1.01	1%	1.2	1.24
251+	--	--	--	--	--	--	--	--	--
Total		3,060			1.35	1.37	5%	1.0	1.30

Appendix G. Critical Items from the Questionnaire

National Survey of the Mining Population.

Final List of Critical Items
October 31, 2008

Question Number	Variable Name
M1a	EMP_TRAIN_REF
M1b	EMP_TRAIN_INEXP
M1c	EMP_TRAIN_EXP
M10	LANG_NON_ENG
M11	MATS_NON_ENG
M14	ADD_MATS_NON_ENG
M16aa	PROD_WORKERS SCH_DAYS_PROD
M16ab	SCH_HOURS_PROD
M16ac	ACT_HOURS_PROD
M16ad	CH_SHIFTS_PROD
M16ae	TRAV_HOURS_PROD TRAV_MINUTES_PROD
M16ba	PROD_SUP_WORKERS SCH_DAYS_PROD_SUP
M16bb	SCH_HOURS_PROD_SUP
M16bc	ACT_HOURS_PROD_SUP
M16bd	CH_SHIFTS_PROD_SUP
M16be	TRAV_HOURS_PROD_SUP TRAV_MINUTES_PROD_SUP
M16ca	PREP_WORKERS SCH_DAYS_PREP
M16cb	SCH_HOURS_PREP
M16cc	ACT_HOURS_PREP
M17a	PROD_WORKERS SHIFTS_DAY_PROD

Question Number	Variable Name
M17b	PROD_SUP_WORKERS SHIFTS_DAY_PROD_SUP
M17c	PREP_WORKERS SHIFTS_DAY_PREP
M23a	USE_CONT_DEVELOP
M23b	USE_CONT_CONST
M23c	USE_CONT_DEMO
M23d	USE_CONT_DAMS
M23e	USE_CONT_EXCAV
M23f	USE_CONT_EQUIP
M23g	USE_CONT_EQUIP_SRV
M23h	USE_CONT_MATERIAL
M23i	USE_CONT_DRILL
M23j	USE_CONT_PROD
M23k	USE_CONT_MINERAL
M23l	USE_CONT_OTHER
M24a	NUM_CONT_DEVELOP
M24b	NUM_CONT_CONST
M24c	NUM_CONT_DEMO
M24d	NUM_CONT_DAMS
M24e	NUM_CONT_EXCAV
M24f	NUM_CONT_EQUIP
M24g	NUM_CONT_EQUIP_SRV
M24h	NUM_CONT_MATERIAL
M24i	NUM_CONT_DRILL
M24j	NUM_CONT_PROD
M24k	NUM_CONT_MINERAL
M24l	NUM_CONT_OTHER
Step3	TOTAL_NUMBER

Question Number	Variable Name
E2	JOB_TITLE
E3,	TITLE_EXP_MNTHS TITLE_EXP_YRS
E4, or	THIS_MINE_MNTHS THIS_MINE_YRS
E5	TOTAL_MINE_MNTHS TOTAL_MINE_YRS
E7	WORK_LOCATE
E8	GENDER
E9 or	LATINO
E10	RACE_INDIAN RACE_ASIAN RACE_BLACK RACE_HAWAIIAN RACE_WHITE

Appendix H. Job Titles as Submitted by Survey Respondents

2nd Floor Operator (Froth Cells)
3rd Stationary Equipment Floor SEO Operator
777 Operator
7820 Operator
ABMO Operator
AC Mill/Screen 3 Operator
Accountant
Accountant 3 Mine Ops
Accountant Operations Technician
Accounting AP-AR
Accounting Assistant
Accounting Associate
Accounting Associate, Senior
Accounting Clerk
Accounting Clerk & HR
Accounting Coordinator
Accounting Manager
Accounting Specialist
Acting CCD Manager
Additive Press Operator
Additives Utilityman
Administration
Administration Accounting
Administration Assistant
Administration Technician
Administrative
Administrative Assistant
Administrative Assistant I
Administrative Assistant Coordinator
Administrative Clerk
Administrative Coordinator
Administrative Lead Man
Administrative Manager
Administrative Office Plant 3
Administrative Office Plant 4
Administrative Secretary
Administrative Services Manager
Administrative Specialist
Administrative Superintendent
Administrative Support

Administrative Technician
Administrator
Administrator III
Advanced Operator
Aggregate Area Manager
Aggregate Plant Laborer
Aggregate Plant Mechanic
Aggregate Welder
A-Helper
Airplane Pilot
A/P Clerk
Apprentice
A/R/Dispatch
Area Leader
Area Manager
Area Production Manager
Assay Lab Technician V
Assayer
Assistant Accounts Payable
Assistant Aggregate Mechanic
Assistant Asphalt Plant Operator
Assistant Belt Coordinator
Assistant Filter Evaporation Operator
Assistant Foreman
Assistant Manager
Assistant Mechanic
Assistant Mine Foreman
Assistant Mine Supervisor
Assistant Office Manager
Assistant Operation Manager
Assistant Plant Manager
Assistant Plant Operator
Assistant Preventative Maintenance Engineer
Assistant Shift Manager
Assistant Superintendent
Assistant Supervisor
Auger Crew Operator
Auger Crew Supervisor
Auger Operator
Auto Bagger

Automation Engineer
Automation Process Engineer
Automotive Mechanic/Standard
Automotive Repairman
Automotive Serviceman—Hostler
Backhoe Operator
Bag Crew
Bag Handler
Bagged Car Loader
Bagger
Bagger 50-lb
Bagger/Labor
Bagger/Operator
Bagger/Sealer
Bagging and Quality Control
Bagging/Loading
Bagging Facility Foreman
Bagging Laborer
Bagging Operator
Bagging Operator Supervisor
Bagging Supervisor
Baghouse Supervisor
Ball Mill Operator
Barge Controller
Barge Loader
Barge Tender
Batch Plant Operator
Bathhouse/Yard
Belo Man
Belt Attendant
Belt Cleaner
Belt Crew
Belt Crew Foreman
Belt Examiner
Belt Foreman
Belt Maintenance
Belt Man
Belt Mechanic
Belt Operator
Belt Patrolman
Belt Piler
Belt Press Operator
Belt Repairman

Belt Tender
Belt Walker Examiner
Belt Worker
Belts
Belts Electrician
Belts General Labor
Beneficiator
Benefits Administrator
Big Bagger Lead Operator
Bin Puller
Bin Tender
Bin Truck Driver
Blade Operator
Blaster
Blaster Helper
Blaster/Primary Operator
Blasting
Blasting Assistant
Blasting Supervisor
Block Press Operator
Block Sawyer
Blunging Operator
Blunging Operator 1
Blunging Operator 4
Boat Operator
Boat Pilot
Bob Cat Operator
Bobcat
Bobcat Operator
Bobcat & Stone Cutter
Boiler/Coating Operator
Boiler Operator
Boiler Plant Operator
Boilermaker
Boilermaker/Welder
Bolter
Bolter Operator
Bookkeeper
Bookkeeper/Accounts Manager
Boss
Bratticeman
Breaker Operator
Bridge Operator

Buggy Runner
Buggy/Shuttle Car
Bulk Loader
Bull Cook
Bulldozer Operator
Burner
Burner Operator
Business Manager
Buyer
Buyer/Coordinator II
Buying Associate
Calcine Big Bagger Lead Operator
Calcine Operator
Calcine Operator 1
Calciner
Car Operator
Carpenter
Carpenter/Painter
Cat Operator
Cedar Rapids Operator
Cement Regional Sales Manager
Central Control Operator
CEO
CEO/President
Certified Welder
Chief Chemist
Chief Clerk
Chief Electrician
Chief Executive Officer
Chief Mechanic
Chief Mechanic/Electrician
Chief Metallurgist
Chief Mine Engineer
Chute Puller
CKD Dust Truck Operator
Classification/Operator Sandgrinder
Clay Operator
Cleanup Man
Clerk
Clerk II
Clerk Scale
Clerk/Scale I
Clerk Scale III

CM Operator
CM Unit Operator IV
CMMS Planner
Coal Cleaner
Coal Distribution Coordinator
Coal Handling Manager
Coal Hauler
Coal Hauler Operator
Coal Miner
Coal Sampler
Coal Testing
Coal Unloader
Communications Supervisor
Concentrator Supervisor
Concrete Man
Console Operator
Construction Crew
Construction Equipment Operator
Continuous Miner
Continuous Miner Operator
Control Analyst
Control Person
Control Room
Control Room Electrician
Control Room Operator
Control Room Supervisor
Controller
Conveyor Man
Conveyor Operator
Conveyor Technician
Cook
Co-op Student
Coordinator
Coordinator Financial Reporting and
    Controls
Cost Coordinator
Cowles Operator
Craftsman A
Craftsman C
Crane Operator
Crane Operator/Truck Driver
Crew Foreman
Crew Leader—Surface

Crude Clay Controller
Crude Lab Technician
Crude Ore Loader
Crude Pile Operator
Crude Prep
Crush & Convey Mechanic
Crush Operator
Crusher
Crusher Attendant
Crusher Foreman
Crusher Foreperson
Crusher Helper
Crusher Helper III
Crusher Laborer
Crusher/Loader Operator
Crusher Maintenance
Crusher Man
Crusher Operator
Crusher Operator Technician V
Crusher Plant
Crusher Plant Operator
Crusher Repairman
Crusher Rock Loader
Crusher Supervisor
Crusher Technician
Crusher—Telsmith Operator
Crusher Utility Person
Crusher Worker
Crushing Foreman
Crushing Leader
Crushing Plant Loader
Crushing Plant Operator
Crushing Supervisor
CS
Ctrl/Electrical Systems Integrator
Curb Cutter
Curb Shed Foreman/Curb Cutter
Curtain Man
Customer Loader
Customer Loader Operator
Customer Service
Customer Service Manager
Customer Service Representative
Customer Service—SF
Cutstone A
Cutstone B
Cutter Operator
D-10 Dozer
Data Processor
Deck Hand
Degritter Operator
Delivery Driver
Demurrage Clerk
Department Helper
Diagnostic Mechanic
Diamond Drill Lead Man
Diamond Sawyer
Diesel Mechanic
Digestion Operator
Director Environmental Services
Director of Coal Sales
Director of Engineering
Director of Scheduling & Logistics
Dispatcher
Dispatcher II
Dispatcher Assistant
Dispatcher/Weighman
Distribution Coordinator
Distribution Manager
Dock Hand
Dock Man
Dock Worker
Dozer
Dozer Driver
Dozer, Excavator, Operator
Dozer/Hilift Operator
Dozer Man/Haul Truck Driver
Dozer Operator
Dozer Operator 1
Draftsman
Dragline
Dragline Assistant
Dragline Oiler
Dragline Operator
Dragline Technician
Dredge & Dozer

Dredge Manager
Dredge Operator
Dredge/Plant Operator
Dredger
Drill/Blast Supervisor
Drill Mucker
Drill Operator
Drill Rig Operator
Driller
Driller III
Driller Blaster
Driller/Miscellaneous
Drilling
Drilling/Blaster Leader
Driver
Driver/Equipment Operator
Driver Haul Truck
Driver Haul Truck I
Driver Haul Truck II
Driver Off Road Truck
Driver/Operator
Driver/Shop Work
Driver Stockpile Truck
Dry Attendant
Dry Plant Lead Man
Dry Plant Manager
Dry Plant Operator
Dry Plant Sacker Operator
Dry Plant Worker
Dry Process
Dry Section Operator
Dryer
Dryer/Loader Operator
Dryer Operator
Drymill Operator
Dump Truck Driver
Dump Truck Operator
Earth Strip
EHS Coordinator
EHS Coordinator Customer Service
EHS Manager
EHS Technician

E/I Technician IV
Electrical Apprentice
Electrical Control Technician
Electrical Coordinator
Electrical Department Coordinator
Electrical Engineer
Electrical Foreman
Electrical/Instrumentation Apprentice
Electrical/Instrumentation Coordinator
Electrical Maintenance
Electrical Maintenance Level C
Electrical Resource
Electrical Supervisor
Electrical Technician
Electrical Technician I
Electrician
Electrician I
Electrician II
Electrician III
Electrician IV
Electrician A
Electrician H
Electrician/Maintenance
Electrician/Maintenance Supervisor
Electrician/Mechanic
Electrician Mine
Electrician STD
Electrician Technician
Electrician Trainee
Electro-Instrumentist
Electronic Repairman
Electronic Technician
Electronic Technician—Standard
Electrowinner
Emergency Response Coordinator
End Dump
End Dump Driver
End Dump Operator
End Loader Operator
Engineer
Engineer II
Engineer Analyst Senior
Engineer/Operations Manager

Engineer Plant Operator
Engineering Intern
Engineering Manager
Engineering Supervisor
Engineering Technician
Entry Bagger
Entry Level Miner
Environmental Engineer
Environmental Engineer I
Environmental, Health & Safety Coordinator
Environmental Health & Safety Manager & RSO
Environmental Manager
Environmental Officer
Environmental Specialist
Environmental Staff
Environmental Technician
E.O. Utility
Equipment Maintenance Manager
Equipment Management
Equipment Manager
Equipment Mechanic
Equipment Mechanic/Fueler
Equipment Oiler
Equipment Operator
Equipment Operator I—SF
Equipment Operator II—SF
Equipment Operator III
Equipment Operator III—SF
Equipment Operator IV
Equipment Operator V
Equipment Operator VI
Equipment Operator/Laborer
Equipment Operator/Manager
Equipment Operator—Material Supplier
Equipment Operator/Mechanic
Equipment Operator (mobile)
Equipment Operator/Supervisor
Equipment Operator—Surface
Equipment/Plant Operator
Equipment Relief
Equipment/Shift Manager
Equipment Trainer
Equipment Training Supervisor
ER Plant
Euclid
Euclid Operator
Evening Dozer/Loader Operator
Evening Driller
Evening Loader Operator
EW Operator
Examiner
Excavator
Excavator Operator
Executive Assistant
Executive Assistant to President
Exploration Driller
Explosives Loader
Explosives Specialist
Explosives Technician
Extruder Operator
Fabricator
Face Boss
Face Driller
Face Loader
Face Loader Operator
Face Man
Face Operative
Facility Manager
Facility Operator
Facility Service Maintenance I
FEL wa 800
Field Electrical Repairman
Field Loader
Field Mechanic
Field Supervisor
Filter Operator
Fine Grind—Surface Plant Manager
Finish End Plant Trainee
Finish Grinder Operator
Fire Boss
Fire Boss/Belt Man
Fire Boss Pumper
Fire Equipment SV
First Line Supervisor

Fixed Equipment Maintenance
Fixed Main Supervisor
Fixed Maintenance I
Flagman
Floating Utility
Flock
Flotation Operator
Flotation Plant Operator
Fluid Bed Dryer Operator
Foreman
Foreman 1st Shift
Foreman 2nd Shift
Foreman/Dredge Operator
Foreman Maintenance
Foreman/Manager/Staff
Foreman/Miner
Foreman Operator
Foreman/Operator
Foreman—Quarry
Foreman Scoop & Buggy Man
Foreman Trainee
Fork Truck Operator
Forklift
Forklift Operator
Forklift Operator & Utility
Front End Loader Operator
Froth Cell Operator
Fuel Electrician
Fuel & Lube Truck Operator
Fuel Mechanic
Fuel Mechanic Helper
Fuel Oiler
Fuel Operator
Fuel Technician
Fuel Truck
Fueler
Gantry Crane Operator
Garage/Machine Shop Maintenance Group
Gate Keeper
General Foreman
General Inside
General Inside Laborer
General Inside/Roof Bolter
General Labor & Equipment Operator
General Labor/Shop Work
General Laborer
General Maintenance
General Manager
General Mine Foreman
General Mine Manager
General Miner Support
General Operation Manager
General Outside
General Outside Laborer
General Plant Helper
General Repairer
General Superintendent
General Supervisor
General Underground
General Underground Laborer
General Utility
Geo Technician II
Geologist
Geologist II
Geophysicist
Gold House Supervisor
Gradall Operator
Grader Operator
Granule Superintendent
Gravel Pumper
Gravity Mag Operator
Greaser
Greaser & Fueler
Greaser/Oiler
Grinder Operator
Grinding
Grinding Float
Ground Control Technician
Ground Hand
Ground Man
Grounds Keeper
Group Leader
Group Leader Ground Packaging
Group Leader Milling
Grouter

Guard, Security II
Gyp Mine Manager
Hammer Operator
Haul Truck
Haul Truck Driver
Haul Truck Driver I
Haul Truck Driver II
Haul Truck Driver—Off Road
Haul Truck Driver—On Road
Haul Truck/Loader
Haul Truck Operator
Haul Unit Driver
Haul Unit Operator
Haul Unit Operator/Stock
Haulage
Haulage Driver
Haulage Operator
Hauler
Hauler Operator
Hauling
HDR
Head Blaster
Head Operator
Heading Prep
Health & Safety Manager
Health & Safety Officer
Health & Safety Technician
Heap Leach Operator
Heavy Duty Mechanic
Heavy Duty Mechanic Welder
Heavy Duty Repair
Heavy Duty Repair Trainee
Heavy Equipment Electrician
Heavy Equipment Mechanic
Heavy Equipment Mechanical Electrician
Heavy Equipment Operator
Heavy Equipment Operator B
Heavy Equipment Operator (Dozer)
Heavy Equipment Operator—Front End Loader
Heavy Equipment Operator—Haul Truck
Heavy Equipment Operator—Lead
Heavy Equipment Operator—Scrapers
Heavy Equipment Operator—Water Truck
Heavy Equipment Repair MT III
Heavy Equipment Repairman
Heavy Mechanic
Heavy Media Plant Operator
Helper
Helper/Laborer
High Lift Loader Operator
High Lift Operator
High Scaler
Highwall Drill Operator
HMS Operator
Hoe Operator
Hoist Engineer
Hoist Operator
Hoisting Engineer
Hoistman
Hopper Operator
Hot Plant Operator/Loader Operator
HR Generalist II
HR Manager
Human Resources
Human Resources/Accounts Receivable
Human Resources Area Manager
Human Resources Assistant
Human Resources Intern
Human Resources Manager
Human Resources Specialist
Hydraulic Scaler
Hydromet Helper
I & C Technician
Idle Work
Industrial Diagnostic Electrician
Industrial Electrician
Industrial Maintenances Technician IA
Industrial Maintenances Technician II
Industrial Plant Bagging
Information Technology Coordinator
Inglett Bagger
Inglett Operator
Inspector
Instrument Electrician

Instrument Repair
Instrument Repairer
Instrumentation Supervisor
Intern
Intern Student
IT Support
IT Technician
Janitor
Janitor/Utility
Jaw Operator
Jet Mill Operator
Jig Plant Operator
Journeyman
Journeyman Electrician
Jumbo Driller
Junior Geologist
Kiln Assistant
Kiln Burner
Kiln Feed Operator
Kiln Laborer
Kiln Operator
Lab
Lab Analyst
Lab Assistant
Lab Chemist
Lab Clerk
Lab Electrician
Lab Manager
Lab Operator
Lab Person
Lab Supervisor
Lab Systems Technician
Lab Technician
Lab Technician I
Lab Technician II
Lab Technician III
Lab Tester
Lab Worker
Labor Foreman
Labor Pool
Laboratory Technician
Laborer
Laborer II—SF

Laborer Equipment Operator
Laborer/Ground Person
Laborer Helper
Laborer/Maintenance
Laborer/Plant Operator
Laborer Roof Slate
Laborer—Pit 2
Laborer/Site Manager
Laborer (Summer)
Laborer—Utility
Lamp Man
Land Manager
Large Shovel/Backhoe/Load Operator II
Large Truck Driver
Leach Pad Operator I
Leach/Roast Operations Helper
Leach Utilityman
Lead Bagger
Lead Electrician
Lead Equipment Mechanic
Lead Laborer
Lead Man
Lead Man—Mill
Lead Man—Mine
Lead Man—Quarry
Lead Man Roller Mill Plant Operator
Lead Man—SF
Lead Man Wet Process
Lead Mechanic
Lead Miner
Lead Operator Mill
Lead Operator Quarry
Lead Payloader
Lead Person
Lead Person II
Lead Plant Operator
Lead Primary Mobile
Lead Process Operator
Lead Warehouse
Ledge Foreman
Ledge Worker
Leech Pad Operator
Level A Certified Blaster

Level A Chief Op or PSO
Level A Millwright 1C
Level B Millwright 2C
Level B Miner
Level C Mine Helper plus Truck
Level C Supply Specialist
Level D Entry
LHD Operator
Lift Driver Lead
Light Vehicle Mechanic II
Limestone Prep Operator
Line 2 Loadout Operator
Line Leader
Liquid Fuel Handler
Load Explosives
Load Out Operator
Loader
Loader Crusher Operator II
Loader/Excavator Operator
Loader/Ground Bagger
Loader Man/Driller
Loader Mine
Loader/Miner
Loader Operator
Loader Operator—Feeds Crusher
Loader Operator—Loads Trucks
Loader Operator Supervisor
Loader Operator—Truck Driver
Loader (Portable)
Loader - Setter
Loader/Stock Truck
Loader, Stockpile
Loader (Yard)
Loadhouse Supervisor
Loading Equipment Operator
Loading Hauler Trucks
Loading Rock in Process
Loading Trucks
Loading & Warehouse
Loadman
Loadout
Loadout Operator
Locomotive Engineer

Longwall Area Manager
Longwall Foreman
Longwall Mechanic Operator Helper
Longwall Production Operator
Longwall Propman
Longwall Shearer Operator
Longwall Support
Longwall Trainee
Lube Bay Oiler
Lube Maintenance
Lube Man
Lube Specialist
Lube Technician
Lube Truck
Luber
Luber—Fixed Equipment
Lubrication Maintenance
Lubrication Repairman
Lubricator
LWDF Attendant
M.E.O.
Machine Loader Operator
Machine Operator
Machinist
Main/Truck Driver
Maintenance
Maintenance V
Maintenance A Electrician
Maintenance A/Utility Leader
Maintenance B
Maintenance Chief
Maintenance Clerk
Maintenance Coordinator
Maintenance Craft
Maintenance Crew
Maintenance Electrician
Maintenance Electrician II
Maintenance Engineer
Maintenance/Equipment
Maintenance Fixed I
Maintenance Fixed II
Maintenance Fixed III
Maintenance Foreman

Maintenance/General Supervisor
Maintenance Group
Maintenance Group Lead Man
Maintenance Helper
Maintenance Inspector
Maintenance Inst.
Maintenance Journey
Maintenance Lead Man
Maintenance Lead Person
Maintenance Leader
Maintenance Level C
Maintenance/Loader Operator
Maintenance Lube
Maintenance/Machine Shop Supervisor
Maintenance Man
Maintenance Man Level A-1
Maintenance Man Machine Lube
Maintenance Manager
Maintenance Mechanic
Maintenance Mechanic I
Maintenance Mechanic II
Maintenance Mechanic lll
Maintenance Mechanic—Standard
Maintenance Mechanic Supervisor
Maintenance Mobile I
Maintenance Mobile II
Maintenance Mobile III
Maintenance/Off-road Truck Driver
Maintenance Operator
Maintenance Planner
Maintenance Planner II
Maintenance Planner/Mechanic
Maintenance/Plant Operator
Maintenance/Plant Supervisor
Maintenance Repairman
Maintenance Superintendent
Maintenance Supervisor
Maintenance System Site Administrator
Maintenance Systems Administration
Maintenance Team Leader
Maintenance Team Member
Maintenance Technician
Maintenance Technician I

Maintenance Technician II
Maintenance Technician Senior
Maintenance Trainee
Maintenance Welder
Maintenance Worker
Maintenances Facilities Technician 1A
Maintenances Facilities Technician B
Maintenances Supervisor
Makedown Technician
Management
Manager
Manager Assistant Plant 2
Manager/Global Screening
Manager—New Polymer Composites
Manager of Administration
Manager of Concentrator
Manager of Engineering
Manager of Financial Reporting
Manager/Owner/Equipment Operator
Manager Plant 3
Manager Trainee
Manager Transmission/Sales
Manager/Vice President
Managerial
Manager's Assistant
Manufacturing Supervisor
Mark Up/Layout
Marketing Services Director
Mass Excavator 5130 cat
Master Electrician
Master Heavy Equipment Operator
Master Mechanic
Master Mill Technician
Master Process Operator
Material Handler
Material Handler II
Material Operator
Material Sampler
Material Unloader
Materials Coordinator
Materials Operator
Materials Planner
Materials & Planning Manager

Materials Technician
MBC Operator
Mechanic
Mechanic I
Mechanic II
Mechanic V
Mechanic A
Mechanic B
Mechanic B—Group Leader
Mechanic/Chief
Mechanic Clerk
Mechanic D
Mechanic Electrician
Mechanic/Electrician
Mechanic/Electrician II
Mechanic/Fabricator
Mechanic G
Mechanic Helper
Mechanic Lead Person
Mechanic Level IV
Mechanic Level V
Mechanic/Maintenance
Mechanic Mobile
Mechanic/Operator
Mechanic—Plant
Mechanic Specialist
Mechanic Technician II
Mechanic Technician III
Mechanic Technician IV
Mechanic Trainee
Mechanic—Truck Driver
Mechanic—Underground
Mechanic Utility
Mechanic/Welder
Mechanical Engineer
Mechanical Engineer/EMR
Mechanical Maintenance
Mechanical Maintenance A
Mechanical Repairman
Mechanical Scaler Operator
Mechanical Technician
Mechanics Helper
Mechanics Helper—Lead
Mechanics Welder
Messenger
Met Lab Technician VII
Metallurgist
Metallurgist II
Mill
Mill Crusher Operator
Mill E&I Technician
Mill Foreman
Mill Hand
Mill Helper
Mill Kiln Operator
Mill Lead Man
Mill Lead Technician IV
Mill Maintenance
Mill Maintenance Technician
Mill Manager
Mill Mechanic
Mill Mechanic Foreman
Mill Mechanic Technician II
Mill Operations
Mill Operator
Mill Operator/Lead Man
Mill/Packaging Operator 1
Mill/Packaging Operator 2
Mill/Packaging Operator 3
Mill Production
Mill Production Laborer
Mill Production Worker
Mill Superintendent
Mill Technician
Mill Technician II
Mill Technician IV
Mill Utility
Mill/Warehouse Operator
Miller
Millerman 1
Millerman 1—Lead Man
Milling Lead Man
Milling Machine Operator
Millman
Millman's Helper
Millwright

Millwright I
Millwright IV
Millwright STD
Millwright STR
Mine A
Mine Apprentice
Mine Clerk
Mine Driller
Mine Electrician
Mine Engineer
Mine Equipment Operator
Mine Examiner
Mine Foreman
Mine Foreman—Miner Operator
Mine Foreman/Superintendent
Mine General Foreman
Mine Haul Truck Driver
Mine—Haul Truck Driver
Mine Hauler
Mine Labor
Mine Lead
Mine Lead Man
Mine Lead Technician IV
Mine Leader
Mine Loader
Mine Loader Operator
Mine Maintenance
Mine Maintenance Clerk
Mine Maintenance Foreman
Mine Maintenance Mechanic
Mine Maintenance MT 3
Mine Maintenance Production Supervisor
Mine Maintenance Specialist
Mine/Maintenance Superintendent
Mine Maintenance Technician II
Mine Maintenance Technician V
Mine Manager
Mine Mechanic
Mine Mechanic I
Mine Mechanic II
Mine Mechanic III
Mine Mechanic A
Mine Oiler/Fueler I

Mine Operations
Mine Operations I
Mine Operations—Equipment Operator
Mine Operations Technician I
Mine Operations Technician IV
Mine Operations Technician V
Mine Operator
Mine Operator C
Mine Production
Mine Production—Hoist operator
Mine Production Operator
Mine Production Superintendent
Mine Production Supervisor
Mine & Quarry Maintenance
Mine & Quarry Manager
Mine Relief Utility
Mine Shift Supervisor
Mine Spec I
Mine Spec II
Mine Spec III
Mine Superintendent
Mine Supervisor
Mine Supplier
Mine Support
Mine Surveyor
Mine Technician III
Mine Technician IV
Mine Truck Driver
Mine—Truck Driver
Mine Utility
Mine Utility B
Mine Worker
Miner
Miner 1
Miner 1st Class
Miner 2
Miner 3
Miner 4
Miner 5
Miner Helper
Miner Lead Man
Miner Operator
Miner Section Operator

Mines
Mining
Mining Engineer
Mining Lead Man
Mining Supervisor
Miscellaneous Operator
Mix Chemist
Mix Control Chemist
Mix Control Fill-in
Mix Man
Mixer
Mixer Operator
Mobile Bridge Operator
Mobile Equipment
Mobile Equipment Maintenance
Mobile Equipment Mechanic
Mobile Equipment Mechanic STD
Mobile Equipment Operator
Mobile Maintenance
Mobile Maintenance Foreman
Mobile Maintenance Mechanic
Mobile Mechanic
Mobile Repair
Mobile Utility Operator
Motor Grader 873 JD
Motor Grader Operator
Motor Grader Operator—Lead
Motorman
Mucker
Mud Picker
Multi Craft Maintenance
Nashtec Operator
Net Work Coordinator
Night Foreman/Evening Dozer
Night Lead Man
Night Mechanic
Night Supervisor
Night Watchman
Nipper
Off Road Truck Driver
Off Road Truck Operator
Off Sider
Office Administration

Office Administrator
Office Assistant
Office Attendant
Office Clerk
Office Coordinator
Office Manager
Office Manager Loader Operator/Scale
Office Salesman
Office Scale
Office/Scale
Office Staff
Office Staff 1
Office Staff 2
Office Staff 3
Oil Helper
Oil Pit Technician
Oiler
Oiler/Maintenance
Oiler/Repairman
On Road Truck Driver/Loader Operator
Op. Tech. Pel
Open Pit 1
Operating Engineer
Operations
Operations Administrator
Operations Associate
Operations Engineer
Operations Engineer/Labor Engineer
Operations Maintenance Technician
Operations Manager
Operations Specialist
Operations Superintendent
Operations Supervisor
Operations Support Clerk
Operations Support Coordinator
Operations Technician Filter Attendant
Operations Technician Material Handler
Operations Technician Prim. Cr. Attd
Operations Technician/Shovel Operator
Operator
Operator I
Operator II
Operator III

Operator III Utility
Operator IV
Operator V
Operator A
Operator A Prime Leader
Operator Apprentice
Operator B
Operator B—Heavy Equipment Operator
Operator C
Operator CM
Operator/CM
Operator D 6
Operator D Utility Equipment Operator
Operator/Dozer
Operator/Driver
Operator Equipment I
Operator Equipment II
Operator Equipment III
Operator Equipment IV
Operator Equipment V
Operator (Extra)
Operator Foreman
Operator/Ground Person
Operator In Charge
Operator K
Operator Loader
Operator/Loader
Operator Maintenance
Operator/Maintenance
Operator/Maintenance Laborer
Operator Maintenance Man
Operator/Mechanic
Operator—Mobile
Operator Plant 2
Operator Plant 4
Operator/Repairman
Operator—Scoop
Operator/Shovel
Operator/Supervisor
Operator Supervisor
Operator Technician II
Operator Trainee
Operator/Truck Driver

Operator—Underground
Order Processor
Ore Technician
Ore Truck
Ore Truck 77D
OTR Truck Driver
Outby
Outby Electrician
Outby Foreman
Outby General Laborer
Outby Labor
Outby Support
Outby Support UG
Outside
Outside Clerk
Outside Communication
Outside Man
Outside Utility/Clerk
Outside Worker
Outside Yard Man
Over the Road Truck Driver
Overburden Driller
Owner
Owner/Manager
Owner/Miner
Owner Operator
Owner/Operator
Owner/Partner
Owner/Sales/Shipping
PA
Pack & Ship Lead Man
Packaging
Packaging/Blending
Packaging Operator
Packaging Supervisor
Packaging Team Member
Packer
Packer Crewman
Packer/Forklift
Packer/Loader
Packer Man
Packer Operator
Packer—Pit 2

Packer—SS
Packhouse
Packhouse Utility
Packing Operator
Packing/Shipping Foreman
Palleter
Palletizer/Meo
Pan Operator
Panel Operator
Part Time Laborer
Part Time Shop
Part Time Yard Worker
Parts
Parts Clerk
Parts Coordinator
Parts Runner
Parts Runner/Accounts Payable
Paver Operator
Payables Clerk
Payloader
Payroll
Payroll Assistant
Payroll Clerk
Payroll/Personnel
Pebble Mill Operator
Pellet Plant Technician
Permit Coordinator
Physical Tester
Picker/Laborer
Pinner Operator
Pipe Fitter
Pit and Plant Truck Driver
Pit Foreman
Pit Hauler
Pit Laborer
Pit Lead Man
Pit Loader
Pit Loader Operator
Pit Operator
Pit Superintendent
Pit Supervisor
Pit Truck Driver
Pit Truck Operator

Planner
Planner I
Planner II
Plant
Plant 1 Operator
Plant 2 Operator
Plant 3 Operator
Plant Accountant
Plant Administrator
Plant Attendant
Plant Clerk
Plant Controller
Plant Controlman
Plant Electrician
Plant Engineer
Plant Engineer/HSE
Plant/Equipment Operator
Plant Foreman
Plant Foreman/Loader Man
Plant Foreperson
Plant Generalist
Plant Ground Man
Plant Helper
Plant Laborer
Plant Lead
Plant Leader
Plant Loader
Plant Loader Operator
Plant Maintenance
Plant Maintenance Group
Plant Maintenance Superintendent
Plant Man
Plant Manager
Plant Manager Intern
Plant Manufacturing Supervisor
Plant Mechanic
Plant Mobile Equipment Operator
Plant Office Administration
Plant Office Administrator
Plant Oiler
Plant Operations
Plant Operator
Plant Operator I

Plant Operator I—SF
Plant Operator II—SF
Plant Operator 2A
Plant Operator IV
Plant Operator (Apprentice)
Plant Operator (Beginning Operator)
Plant Operator—Foreman
Plant Operator/Truck Driver
Plant Person
Plant/Pit Foreman
Plant/Pit Truck Driver
Plant Production Worker
Plant Quality & Shipping
Plant Repair
Plant Repair Foreman
Plant Repair/Welder
Plant Repairman
Plant Repairman I
Plant Repairman II
Plant Sampler
Plant Superintendent
Plant Supervisor
Plant Supervisor II
Plant Supervisor Manager
Plant Technician
Plant Technician—Crew Leader
Plant Trainee
Plant Utility
Plant Utility Operator
Plant Wash Operator
Plant Welder
Plant Working Foreman
Plants Manager
Poly Packer Crewman
Polygloss Bagger Technician
Port Operator B
Portable Plant Operator
Powder Loader
Powder Man
Powder Person
Power Screen Operator
Power Systems Operator B
Prep Plant Mechanic

Prep Plant Operator
Prep Plant Technician
President
President/COO
President/Developer/Operator
President/Owner
President/Owner/Retired
Pricing Coordinator
Primary Control Operator
Primary Crusher
Primary Crusher Operator
Primary Mobile Operator
Primary Operator
Primary Operator (Jaw)
Process Assistant
Process Attendant
Process Control Operator
Process Control Superintendent
Process Control Supervisor
Process Control Technician
Process Engineer
Process/Equipment Operator
Process Foreman
Process Lab Technician
Process Laborer
Process Maintenance Mechanic
Process Maintenance Technician IV
Process Maintenance Technician VI
Process Maintenance Utility
Process Manager
Process Operations Technician III
Process Operator
Process Operator II
Process Production Engineer III
Process Supervisor
Process Technician
Processing Assistant 1
Processing Plant
Processing Team Member
Processor
Procurement Manager
Procurement Specialist
Product Loading

Production
Production 1st shift
Production 2nd shift
Production 3rd shift
Production Assistant
Production Coordinator
Production Driller
Production Employee
Production Engineer
Production Expeditor
Production Foreman
Production Generalist
Production Inspector
Production Journeyman
Production Lead Man
Production Lead Operator
Production Leader
Production Loader
Production Loader Operator
Production Maintenance
Production/Maintenance Manager
Production/Maintenance Supervisor
Production Manager
Production Mechanic
Production Miner
Production Operator
Production Operator I
Production Operator II
Production Operator Level I
Production Operator Level II
Production Operator Level III
Production Operator Screening Plant
Production Quality Control Manager
Production Resource Manager
Production & Sales Service Lab Technician
Production Scheduler
Production Scheduler/Safety Manager
Production Shift Foreman
Production Superintendent
Production Supervisor
Production Support
Production Technician
Production Technician II
Production Technician IV
Production Technician V
Production Truck
Production Truck Driver
Production Utility Man
Production Worker
Professional
Project Engineer
Project Manager
Prospecting
Pug Mill Operator
Pug Operator
Pump Man
Pump Operator
Pumper
Purchase Agent
Purchaser
Purchasing
Purchasing Agent
Purchasing Clerk
Purchasing Coordinator
Purchasing Equipment Manager
Purchasing Manager
Purchasing/Shop
Q-line II
Quality Analyst
Quality and Safety Manager
Quality Assistant
Quality Assurance
Quality Assurance Coordinator
Quality Assurance Lab Technician Level II
Quality Assurance Lab Technician Level III
Quality Assurance Manager
Quality Assurance & Mine Supervisor
Quality Assurance/Quality Control Laboratory Technician
Quality Assurance Supervisor
Quality Assurance Technician
Quality Control
Quality Control III Lab Technician

Quality Control/HS&E
Quality Control Lab
Quality Control Lead Technician
Quality Control Man
Quality Control Manager
Quality Control Physical Tester
Quality Control & Sales Coordinator
Quality Control Supervisor
Quality Control Technician
Quality Control Technician II
Quality Loader Operator
Quality Manager
Quality Supervisor
Quality Technician
Quarry Coordinator
Quarry Crusher Operator
Quarry Driller/Blaster
Quarry Equipment Operator
Quarry Extra
Quarry Foreman
Quarry Laborer
Quarry Loader Operator
Quarry Manager
Quarry Mechanic
Quarry Night Foreman
Quarry Operator
Quarry Saw Operator
Quarry Superintendent
Quarry Supervisor
Quarry Technician
Quarry Truck Driver
Quarry Utility
Quarry Worker
Quarryman
Quarryman A
Quarryman B
R&D Supervisor
Rail Lead Man
Rail Loader
Rail Loader Operator
Rail Loadout
Rail Loadout Operator
Rail Operator

Rail Road
Rail Runner
Rail Runner Operator
Rail Supervisor
Raisebore Operator
Rak Handler
Ram Car Operator
Raw Material Manager
Raymond Mill Operator
Receiving Clerk
Receiving Supervisor
Receptionist
Receptionist/Shipping Coordinator
Reclaim Operator
Reclamation Labor
Refuse Site Operator
Refuse Truck Operator
Regional Human Resources Manager—U.S.
Regulatory Manager
Reliability Engineer
Repair Lead Man
Repair Worker
Repairman
Repairman A
Representative Trade Relations
Research Scientist
Road Grader Operator
Road Maintenance
Robot Operator
Rock Breaker
Rock Breaker Operator
Rock Crusher Superintendent
Rock Duster
Rock Haul Driver
Rock Plant Operator
Rock Truck
Rock Truck/Dozer Operator
Rock Truck Driver
Rock Truck Operator
Roller Mill Operator
Roller Mill Plant Operator Fine Grind
Roller Operator

Rolling Stock Crew 2
Rolling Stock Crew 4
ROM Operator
Roof Bolter
Roof Bolter Operator
Roof Bolter—Scaler
Roof Control
Roof Control Operator
Roof Drill
Roof Person
Roofing Slate Splitter
Roofing Slate Trimmer
Roofing Slate Trimming Machine Operator
Root Picker
Rotary Drill Operator
Rotary Dump Operator
Rotex Operator
Roustabout
Roving Clerk
RP Operator
Rubber Tire Operator
Sacker
Sacking
Safety
Safety Advisor
Safety Clerk
Safety Coordinator
Safety Director
Safety Engineer
Safety & Health
Safety & Health Professional
Safety/HR Manager
Safety & Inventory Coordinator
Safety Manager
Safety Officer
Safety Representative
Safety/Security Director
Safety Specialist
Safety Supervisor
Safety Technician
Sales
Sales Administration Manager
Sales Coordinator
Sales Loader
Sales Manager
Sales & Marketing
Sales Person
Sales Representative
Sales Representative 1
Sales/Safety Director
Sales & Technical Manager
Sales/Traffic
Salesman
Salesman Manager
Sample Prep
Sampler
Sampler—Lab
Sampler Technician
SAMS Technician
Sand Plant Lead Man
Sand Plant Operator
Saw
Saw & Equipment Repair
Saw Operator
Saw & Stone Cutter
Saw Table Laborer
Sawyer
Scale
Scale Attendant
Scale Clerk
Scale House
Scale House Clerk
Scale House Master
Scale House Operator
Scale Man
Scale Master
Scale Office
Scale/Office
Scale Office Dispatcher
Scale Office Manager
Scale Operator
Scale Operator/Office
Scale Operator/Parts
Scale Person
Scale/Sales Office

Scaler
Scaler Operator
Scales/Weights
Scheduler
Scoop
Scoop Loader
Scoop Man
Scoop Operator
Scoop Tractor Operator
Scraper Operator
Screed Person
Screen & Mill Operator
Screen Operator
Screen Plant Labor
Screen Plant Operator
Screenhouse/Crusher
Seasonal Production
Secondary Foreman
Secondary Plant Operator
Secretary
Secretary—Treasurer
Section Boss
Section Electrician
Section Foreman
Section Trainee
Section Trainee IV
Sectional Dock Manager
Security
Security Chief/Safety Trainer
Security Guard
Security Guard/General Laborer
Security Officer
Security Supervisor
Security Watch
Senior Accountant
Senior Accountant II
Senior Accounting Assistant
Senior Accounting Clerk
Senior Administrative Clerk
Senior Controller
Senior Designer
Senior Drafter
Senior Engineer

Senior Geologist
Senior Human Resources Manager
Senior Human Resources Representative
Senior Lab Technician
Senior Lead Plant Operator
Senior Maintenance Mechanic
Senior Maintenance Planner
Senior Maintenance Planner I
Senior Mill Operator
Senior Mine Engineer
Senior Mine Geologist
Senior Mining Engineer
Senior Operator
Senior Operator Maintenance
Senior Planning Clerk
Senior Plant Office Administrator
Senior Plant Operator
Senior Process Control Engineer
Senior Process Control Specialist
Senior Process Controller
Senior Process Operator
Senior Quality Control Technician
Senior Research Technician
Senior Stores Specialist
Senior Vibration Technician
Senior Welder
Service Foreman
Service Man
Service Mechanic
Service Technician
Service Truck Driver
Setup Foreman
Shaft Crew
Shaft Repair
Shearer Operator
Shedder
Shift Foreman
Shift Foreman Mill
Shift Laborer
Shift Maintenance
Shift Manager
Shift Mine Manager
Shift Repairman

Shift Supervisor
Shift Tire Attendant
Shift Utility
Shift Welder Repair A
Shiftbreaker—Lewis
Shiftbreaker—Pit 2
Shifter
Shipping
Shipping Assistant
Shipping Clerk
Shipping Coordinator
Shipping Foreman
Shipping Lead Man
Shipping Loader
Shipping Loader Operator
Shipping Manager
Shipping Operator
Shipping & Receiving
Shipping & Receiving Clerk
Shipping Scales Lead person
Shipping Specialist
Shipping Supervisor
Shipping Team Member
Shipping Technician
Shooter
Shop
Shop/Drag Line
Shop Foreman
Shop Manager
Shop Mechanic
Shop Person
Shop/Plant
Shop Serviceman
Shop Supervisor
Shot Crew
Shot Firer
Shovel Dragline Operator
Shovel Loader Operator
Shovel/Loader Operator
Shovel OB pc1800
Shovel Operator
Shovel Pit Loader Operator
Shuttle Car

Shuttle Car Driver
Shuttle Car Operator
Shuttle Car Operator 21
Silo Operator
Site Mechanic/Welder I
Site Superintendent
Skid Steer Operator
Skilled Instrument Electrician 1C
Skilled Laborer
Skilled Maintenance Mechanic
Skilled Maintenance Worker
Skilled Repairman
Skip Loader
Skip Tender
Slate Carrier
Slate Splitter
Slate Trimmer
Sloop Operator
Slurry Operator
Slurry Operator 1& 2
Slurry Track Technician
Small Bagger Lead Operator
Special Loader
Splitter
Stacker
Stacker Operator
Staff Accounting Specialist
Staff Chemical Engineer
Station Operator
Stationary Equipment Operator
Steamer
Stick Picker
Stock Loader
Stock Out Truck Driver
Stock Pile Driver
Stock Pile Hauler
Stock Pile Loader
Stock Pile Operator
Stock Pile Truck
Stock Pile Truck Driver
Stock Piler
Stock Truck
Stock Truck Driver

Stock Truck/Plant Operator
Stockroom Attendant
Stone Cutter
Stone Cutter, Driver—MAC
Stone Packaging Operator
Stone Splitter
Stone Stacker
Storage Operator
Storeroom
Storeroom Attendant
Storeroom Clerk
Storeroom Floorman
Storeroom Manager
Storeroom Supervisor
Stove Plant Operator
Stripping Dredge Operator
Stripping Operator
Sublevel Miner
Summer Grounds Keeper
Super Sack Operator
Superintendent
Superintendent/Secretary
Superintendent Maintenance
Supervisor
Supervisor 2nd Shift
Supervisor & Backhoe Operator
Supervisor Concentrator
Supervisor Crush/Convey
Supervisor/Dozer Operator
Supervisor Leach Pad
Supervisor Mechanics
Supervisor Mine
Supervisor—Mine
Supervisor Mobile Equipment Quarry
Supervisor —Moly Processing
Supervisor/Operator
Supervisor Plant 1
Supervisor Plant 2
Supervisor/Plant Operator
Supervisor Quality Assurance
Supervisor—Shovel/Drill Maintenance
Supervisor—Tailings
Supervisor Trainee

Supply Clerk
Supply Hauler
Supply Man
Supply/Track Man
Support Foreman
Support Opr. 5
Surface
Surface Coordinator
Surface Electrician
Surface Foreman
Surface General Laborer
Surface Laborer
Surface Maintenance
Surface Maintenance Manager
Surface Maintenance Mechanic
Surface Manager
Surface Mechanic A
Surface Mechanic C
Surface Mine Supervisor
Surface Operations Manager
Surface Operations Technician IV
Surface Operator B
Surface Operator C
Surface Outside
Surface Plant Operator
Surface Production
Surface Production Operator
Surface Production Supervisor
Surface Shift Foreman
Surface Supervisor
Surface Support
Surface Utility
Surface Utility Man
Surveyor
Sweeper Operator
Swingman
SX Helper
SX Operator
System Administrator
Systems Analyst
Tailings Dam Operator
Tailings Foreman
Tailings Pond Operator

Tailings Repairman
Tandem Tractor
Tank Car Washer
Tank Car Washout Technician
Tank House Harvestor
Team Leader
Teamster
Tech II
Tech III
Technical Coordinator
Technical Services Manager
Technical Specialist I
Technician
Technician Quality Control
Technician Quality Control II
Technician Quality Control IV
Technician Quality Control V
Technician Senior
Technologist—Analytical Lab
Temporary Section Foreman
Temporary Worker
Terminal Operator
Thickener Operator
Third Shift Foreman
Tipple Foreman
Tipple Helper
Tipple Hilift Operator
Tipple Operator
Tire Man
Tire Technician
Top Lab Analyst
Top Operator
Tower Cleaner
Tower Operator
Tower Ranger
Track
Track Bolter
Track Driller
Track Foreman
Track Hoe
Track Hoe Operator
Track Man
Track Operator

Tractor Operator
Tractor Operator Loader
Tractor Trailer Driver
Tractor Worker
Trades Person I
Trades Person II
Traffic Coordinator
Traffic Representative
Train Engineer
Train Operator
Trainee
Trainer/Assessor
Trainer Electrician
Transportation Coordinator/Administrative Assistant
Transportation Supervisor
Treasurer
Truck Bin Attendant
Truck Driver
Truck Driver I
Truck Driver II
Truck Driver 50T
Truck Driver/Blaster Helper
Truck Driver Heavy
Truck Driver/Mechanic
Truck Dump Operator
Truck Lead Man
Truck Loader
Truck Maintenance
Truck Operator
TSP General Laborer
TSP Mobile Equipment Operator
TSP Pumper
TSP Worker
Undercutter Operator
Underground Belt Man
Underground Blaster
Underground CM Maintenance Operations
Underground CM Production
Underground CM Set-up
Underground Construction
Underground Construction I

Underground Construction Crew
Underground Electrician
Underground Equipment Operator
Underground Foreman
Underground Laborer
Underground Lead Man
Underground Loader Operator
Underground Manager
Underground Mechanic
Underground Miner
Underground Miner 2/1
Underground Miner 3/1
Underground Miner 3/2
Underground Miner 3/3
Underground Operator
Underground Operator I
Underground Plant Operator
Underground Roof Bolter Operator
Underground Scaler
Underground Shift Foreman
Underground Superintendent
Underground Supervisor
Underground Truck Driver
Underground Utilityman
Unit Helper
Universal Operator
Utility
Utility/Beltline
Utility/Belts
Utility/Bolter
Utility Centrifuge Technician
Utility Engineer Technician
Utility Equipment Operator
Utility Field
Utility Laborer
Utility Lubricator
Utility Man
Utility Man/Surface
Utility Operator
Utility Operator C
Utility Person
Utility Person Field
Utility Person Laborer
Utility Person Pit
Utility Person Plant
Utility Person/Warehouse
Utility Scaler
Utility Technician
Utility Technician Equipment Cleaner
Ventilation
Vertical Driller
Vice President
Vice President Cement Operations
Vice President & General Manager
Vice President/Manager of Aggregate Division
Vice President of Finance/CAO
Vice President/Office Manager
Vice President Sales
Vice President Sales & Marketing
Vice President/Secretary
Vice President Technology
Warehouse
Warehouse 1
Warehouse Coordinator
Warehouse Man
Warehouse Meo
Warehouse Operator
Warehouse Person
Warehouse Supervisor
Warehouse Supervisor/Purchasing Agent
Warehouse Team Leader
Warehouse Technician
Warehouse Worker
Warehouser
Wash Operator
Wash Plant
Wash Plant Operator
Wash Plant Super
Watchman
Water/Sweeper Truck Operator
Water Truck
Water Truck Driver
Water Truck/Fueler
Water Truck Operator
Water Wagon Operator

W'Coat Packer
Weigh Man
Weigh Scale Operator
Weighmaster
Weighmaster/Dispatch
Weld Shop Maintenance Manager
Welder
Welder I
Welder/Fabricator
Welder/Laborer
Welder/Maintenance
Welder/Mechanic
Welder Mill Maintenance
Welder/Pipe Fitter
Welder/Plant Maintenance I
Welder/Plant Maintenance III
Welder/Plant Operator
Welder Repair A
Welder/Repairman
Welder—Standard
Wet Grind Operator

Wet Plant
Wet Plant Attendant
Wet Plant Operator
Wet Process Operator
Wet Utility
Worker
Working Foreman
Working Foreman Loading
Working Foreman Quarry
Wrens Maintenance II
Wrens Maintenance IV
Wrens Maintenance V
Yard
Yard Foreman
Yard Laborer
Yard Loader
Yard Loader Operator
Yard Loaderman (Front End Loaders)
Yard Production Laborer
Yard Truck Driver

Appendix I. Glossary

Unless otherwise noted, the source of the definitions in this Glossary is the Dictionary of Mining, Mineral, and Related Terms [American Geological Institute 1997].

Auger. A rotary drilling device used to drill shot holes or geophone holes in which the cuttings are removed by the device itself without the use of fluids.

Backhoe. A versatile rig used for trenching.

Bagger/bagging operations worker. A worker who typically works at a two or four station filling machine, placing empty bags (generally 50 or 100 lb capacity) on each of the machine's fill nozzles. When each bag is filled, either the filling machine mechanically ejects the bag onto a conveyor, or the operator manually removes the bag and places it on a conveyor or on a pallet for shipping [Cecala and Thimons 1992].

Belt vulcanizer. Equipment that consists essentially of two heavy metal plattens that are placed one on each side of the previously prepared joint and clamped firmly together. Each platten is heated, and this combined application of heat and pressure over a period completes the joint.

Beltman/conveyor man. A worker who sets up and tends chain, belt, or shaker (reciprocating) conveyors to transport coal or metal ore about a tipple at the surface from working the working face in a mine.

Bin puller. A worker who transfers material from a storage bin or chute into mobile equipment for transportation.

Blunging. The process of amalgamating, blending, or beating up or mixing in water.

Bob cat. A miniature front-end loader.

Brattice. A wall or petition in underground mines to control proper circulation of air through work places and passageways. Can be made of wood, canvas, or other materials.

Breakers. A machine used for the primary reduction of coal, ore, or rock [Thrush 1968].

Bull dozer. A tractor on the front end of which is mounted a vertically curved steel blade held at a fixed distance by arms secured on a pivot or shaft near the horizontal center of the tractor. The blade can be lowered or tilted vertically by cables or hydraulic rams. It is a highly versatile piece of earth excavating and moving equipment especially useful in land clearing and leveling work, in stripping topsoil, in road and ramp building, and in floor or bench cleanup and gathering operations. Also called dozer.

Calcine. By heating, to expel volatile matter as carbon dioxide, water, or sulfur, with or without oxidation; to roast; to burn.

Cleanup man. A worker who collects all the valuable product of a given period of operation in a stamp mill, or in a hydraulic or placer mine. Collects and loads spillage resulting from normal operations.

Coal sampler. A worker who cuts a representative part of an ore (or coal) deposit, which should truly represent its average value, and who collects and prepares samples of coal for analysis.

Continuous miner. A mining machine designed to remove coal from the face and to load that coal into cars or conveyors without the use of cutting machines, drills, or explosives.

Controller. Any mechanical or electrical device that is part of or added to a machine or device for automatic regulation or control.

Crude pile. A substance in its natural unprocessed, unrefined state. Crude ore or crude oil, for example. In a natural state; not cooked or prepared by fire or heat; not altered or prepared for use by any process.

Crusher operator/ pan feeder operator. In the mineral and nonmineral industry, including coal, quarry products, mineral and nonmineral ores, a worker who operates a machine that crushes rock or other material and regulates the flow of such material into and from the crusher to the next point of processing or use.

Culm. In anthracite terminology, the waste accumulation of coal, bone, and rock from old dry breakers. In bituminous coal preparation, culm corresponds to slurry or slime, depending upon the size distribution of the suspended solids.

Culm bank. The deposit on the surface of culm usually kept separate from deposits of larger pieces of slate and rock.

Curb. A timber frame, circular or square, wedged in a shaft to make a foundation for walling or tubbing, or to support, with or without other timbering, the walls of the shaft; the heavy frame or sill at the top of a shaft.

Cutting machine. A power-driven machine used to undercut or shear.

Dragline. A type of excavating equipment that casts a rope-hung bucket a considerable distance, collects the dug material by pulling the bucket toward itself on the ground with a second rope, then elevates the bucket, and dumps the material on a spoil bank, in a hopper, or on a pile.

Digestion operator. A worker who tends the battery of digester vessels that dissolve bauxite in plant liquor by: turning valves on pumps to transfer liquid and bauxite slurry through heaters into digester vessels, turning valves to inject milk of lime into vessels, adjusting pumps and valves to circulate cleaning solution through process lines, and collecting samples of slurry and alumina solution for laboratory analysis [DOT 2003].

Dredge. A large floating machine used in underwater excavation for developing and maintaining water depths in canals, rivers, and harbors; raising the level of lowland areas and improving drainage; constructing dams and dikes; removing overburden from submerged ore bodies prior to open pit mining; or recovering subaqueous deposits having commercial value.

Dry plant/dry process. A method of treating ores by heat as in smelting; used in opposition to the wet process.

End dump. Process in which earth is pushed over the edge of a deep fill and allowed to roll down the slope [Infomine Inc. 2010].

Face. The exposed surface of a coal or ore deposit in the working place where mining is proceeding.

Fire boss. A person designated to examine the mine for gas and other dangers usually before but also during the shift. Also known as a mine examiner.

Floatation/concentrator. A plant where ore is separated into values (concentrates) and rejects (tails) or an appliance in such a plant, e.g., flotation cell, jig, electromagnet, shaking table.

Front end loader. A tractor loader with a digging bucket mounted and operated at the front end of the tractor that both digs and dumps in front.

Froth cell. The process for cleaning fine coal, copper, lead, zinc, phosphate, kaolin, etc. with the aid of a reagent; the coal or minerals become attached to air bubbles in a liquid medium and float as a froth.

Geologist. One who studies planet Earth, the materials of which it is made, the processes that act on these materials, the products formed, and the history of the planet and its life forms since its origin.

Grader. A self-propelled or towed machine provided with a row of removing or digging teeth and (behind) a blade to spread and level the material.

Ground control/timberman. A worker who installs timbers in a mine to support the roof and walls of haulage ways, passageways, and the shaft.

Hammer mill. A pulverizing unit consisting of a rotor, fitted with movable hammers that is revolved rapidly in a vertical plane within a closely fitting steel casing. The hammers hit falling rock, which is fractured on impact, or by collision with other rocks or with the casing. When sufficiently reduced in size, the pulverized rock escapes through grids in the casing.

Haulage. The drawing or conveying, in cars or otherwise, or movement of workers, supplies, ore, and waste both underground and on the surface. Generally refers to track mining as opposed to conveyor mining, although belt conveyor systems are sometimes referred to as belt haulage; the system of hauling coal or minerals out of a mine.

Head area. The top portion of a seam in the coal face.

Highwall. The unexcavated face of exposed overburden and coal or ore in an opencast mine or the face or bank on the uphill side of a contour strip mine excavation.

Hoist operator. In mining, a person who operates steam or electric hoisting machinery used to lower cages (elevators) and skips (large, metal, boxlike containers) into a mine and to raise them to the surface from different levels. The worker may be designated according to the type of power used, as an electric-hoist person or steam-hoist person.

Hopper. A storage bin or a funnel that is loaded from the top and discharges through a door or chute in the bottom.

Inby. Toward the working face, or interior, of the mine; away from the shaft or entrance; opposite of outby.

Inspector. One who checks the mine to determine the health and safety conditions. This person makes examinations of and reports on mines and surface plants relative to compliance with mining laws, rules and regulations, safety methods, etc. State inspectors have authority to enforce State laws regulating the working of the mines. Federal inspectors have authority to enforce Federal laws in coal mines.

Jack setter. Miner who assists in the operation of an auger-type underground mining machine; duties include seeing that the roof of the mine at or near the machine is in a safe condition.

Jaw operator. One who operates a machine for reducing the size of materials by impact or crushing between a fixed plate and an oscillating plate, or between two oscillating plates, reducing large rocks, or ores to sizes capable of being handled by any of the secondary crushers.

Kiln. A large furnace used for baking, drying, or burning firebrick or refractories, or for calcining ores or other substances.

Lab technician. One who conducts chemical and physical laboratory tests to assist scientists in making qualitative and quantitative analyses of solids, liquids, and gaseous materials for research and development of new products or processes, quality control, maintenance of environmental standards, and other work involving experimental, theoretical, or practical application of chemistry and related sciences [BLS 2010].

Lampman. In mining, one who cleans, tests, and repairs lamps used underground by miners.

Leaching operator. In ore dressing, smelting, and refining, one who dissolves valuable metal out of ore or slime, using chemical solutions.

Longwall. A long face of coal. A method of working coal seams. The workings advance (or retreat) in a continuous line, which may be several hundred yards in length. The space from which the coal has been removed (the gob, goaf, or waste) which is either allowed to collapse (caving) or is completely or partially filled or stowed with stone and debris.

Metallurgist. One who is skilled in, or who practices, the science and art of separating metals and metallic minerals from their ores by mechanical and chemical processes; one involved in the preparation of metalliferous materials from raw ore.

Mill (rod/ball/pebble). A mineral treatment plant in which crushing, wet grinding, and further treatment of ore is conducted. The plant separates components, such as ball mill, hammer mill, and rod mill that grinds material, with or without liquid, using a rotating cylinder or conical mill, and using balls, rods, or pebbles as grinding material.

Millwright. One who installs, dismantles, or moves machinery and heavy equipment according to layout plans, blueprints, or other drawings [BLS 2010].

Mine examiner. A person designated to examine the mine for gas and other dangers usually before but also during the shift. Also known as a fire boss.

Mobile bridge. A continuous haulage system commonly consisting of an alternating series of piggyback mobile bridge carriers (MBCs) and chain bridge conveyors. They are either physically attached to the continuous miner or detached and independently trammed behind the miner [MSHA 2011].

Mucker. In mining and quarrying, a laborer who shovels ore or rock into mine cars or onto a conveyor from which mine cars are loaded and at some point are removed from the working face or surfaces of natural stone deposits; or one who works in a stope shoveling ore into chutes from which it is loaded into cars on haulage level below.

Open pit. A mining operation designed to extract minerals that lie near the surface. Waste, or overburden, is first removed, and the mineral is broken and loaded, as in a stone quarry.

Outby. Nearer to the shaft, and therefore away from the face, toward the pit bottom or surface; toward the mine entrance. The opposite of inby.

Overburden. Material of any nature, consolidated or unconsolidated, that overlies a deposit of useful materials, ores, or coal, and especially those deposits that are mined from the surface by open cuts.

Palletizer. One who secures battens (grooved strips of wood) around bundles of packaged metal extrusions to form protective shipping pallets, using strapping tool [DOT 2003].

Payloader. Equipment used for excavating.

Pelletizing operations worker. An operator of an apparatus in which finely divided material is formed into small spherical pellets by the use of pressure, centrifugal force, or additives.

Pit. A mine, quarry, or excavation worked by the open-cut method.

Preparation plant. Any facility where coal, or other mixed material, is prepared for market; through common usage, it has come to mean a rather elaborate collection of facilities where mixed material is separated from its impurities, washed and sized, and loaded for shipment.

Pug operator/mixer tender. One who mixes ground preheated magnesia and carbon with hot asphalt in a pug mill to form a viscous mixture suitable for processing into pellets.

Pumper. In bituminous coal mining, a person who works a hand pump to force water, accumulated underground in low places, into a drainage ditch flowing to a natural outlet or pumping station.

Quarry. An open or surface mineral working, usually used for the extraction of building stone, such as slate, limestone, etc. It is distinguished from a mine because a quarry usually is open at the top and front, and, in ordinary use of the term, by the character of the material extracted.

Raise borer. A machine used to produce a circular excavation either between two existing levels in an underground mine or between the surface and an existing level in a mine. In raise boring, a pilot hole is drilled down to the lower level; the drill bit is removed and replaced by a reamer head having a diameter with the same dimension as the desired excavation. This head then is rotated and pulled back up towards the machine.

Reclaim. The process of digging from stockpiles; also, the reprocessing of previously rejected material.

Refuse (pile). Waste material in the raw coal that has been removed in a cleaning or preparation plant; also called tailings.

Rock breaker. A kind of hammer which is used to crush (break) rocks; it is a static piece of equipment; to be operated, it must be attached to another implement [Infomine Inc. 2010].

Rock duster. In bituminous coal mining, a laborer who sprinkles rock dust by hand or with a machine throughout mine workings as a precaution against explosions.

Rolling mill. A rolling mill or establishment for rolling metal into forms.

Roof bolter. In bituminous coal mining, one who reinforces roofs of mine haulage ways, side drifts, and working places with metal or timber to prevent rock and slate falls.

Rotary excavator. Earth-moving machine with a vertical wheel that carries digging buckets peripherally. These loosen soil and deliver to a short conveyor loader, the assembly being mounted on crawler track. Capacity up to 5,000 st/h (4,500 t/h). Also called bucket wheel excavator.

Rotary dump car. A standard small car in which the car body is mounted on a turntable in the car frame. The car body may be swung by hand to dump over either side or either end.

Rubber-tired haulage. The underground use of tractors and dump truck haulage, of the battery or diesel type, and battery-driven shuttle cars.

Safety director. One who promotes worksite or product safety by applying knowledge of industrial processes, mechanics, chemistry, psychology, and industrial health and safety laws [BLS 2010].

Sawyer. In stonework industry, a general term applied to workers engaged in cutting stone with power-driven saws.

Scaler (hand or mechanical). A laborer who knocks the roasted lead ore off grates with a bar as it is dumped from conveyors into cars below, prior to melting, to separate and recover the lead. Lead ore is loaded on grates attached to a conveyor and carried through a furnace in which the sulfur is driven off by roasting.

Scoop car. Diesel or battery-powered equipment with a scoop attachment for cleaning up loose material, for loading mine cars or trucks, and hauling supplies.

Scraper. a. A rod for cleaning out a shothole prior to charging with explosives. b. A mechanical contrivance used at collieries to scrape the culm or slack along a trough to the place of deposit. c. A machine used in mines for loading cars and transporting ore or waste for short distances. There are two basic types of scraper: (1) the hoe or open type, which is particularly suitable for moving coarse, lumpy ore; and (2) the box or closed type, which is particularly suited for handling fine material, especially on a loading slide. d. A digging, hauling, and grading machine having a cutting edge, a carrying bowl, a movable front wall (apron), and a dumping or ejecting mechanism. Also called carrying scraper or pan. e. An apparatus used to take up coal from the floor of a mine, after it has been shot and deposit it either in cars or in a conveyor [Infomine Inc. 2010].

Screed. a. A strip of plaster or wood applied to a surface to be plastered to serve as a guide for making a true surface. b. A wooden strip serving as a guide for making a true level surface on a concrete pavement. c. A board or metal strip dragged across a freshly poured concrete slab to give it its proper level [Dictionary.com 2011].

Screening machine. An apparatus having a shaking, oscillatory, or rotary motion, used for screening or sifting coal, stamped ores, and the like.

Shaft. An excavation of limited area compared with its depth; made for finding or mining ore or coal, raising water, ore, rock, or coal, hoisting and lowering workers and material, or ventilating underground workings. The term is often specifically applied to an approximate vertical shaft, as distinguished from an incline or inclined shaft. A shaft is provided with a hoisting engine at the top for handling workers, rock, and supplies; or it may be used only in connection with pumping or ventilating operations.

Shaft mine. A mine in which the coal seam is reached by a vertical shaft which may vary in depth from less than 100 ft (30 m) to several thousand feet.

Shearer operator. In bituminous coal mining, one who operates a type of coal-cutting machine that shears (cuts) out a channel down the sides of the working face of coal (as distinguished from undercutting) prior to blasting the coal down.

Shooter. One who sets off blasts in a mine or quarry.

Shuttle car operator. One who operates a vehicle on rubber tires or continuous treads to transfer raw materials, such as coal and ore, from loading machines in trackless areas of a mine to the main transportation system.

Skip tender/cager/station attendant. One who directs station operations and movement of cages used to raise and lower workers, mine cars, and supplies between various levels and surface; one who works at the top of a shaft or at an intermediate level inside a mine.

Slurry operator. In ore dressing, smelting, and refining, a laborer who sprays the inner surfaces of furnace walls and roofs with a slurry of silica, water, and fireclay to protect brick, using a compressed-air gun.

Splitter. One who separates blocks of rough dimension stone from quarry mass using jackhammer and wedges [BLS 2010].

Stope. An excavation from which ore has been removed in a series of steps. A variation of steps. This term is usually applied to highly inclined or vertical veins.

Strip. In mining, to remove the earth, rock, and other material from the mineral to be mined, usually by power shovels. Generally practiced only where the mineral lies close to the Earth's surface.

Surface shops. Mining operations do much of their repair work in-house. This work is carried out in shops located on the surface [Vaught 2008].

Surveyor/transit man. One who applies special knowledge and techniques gained through experience or training to make surface and underground surveys at a mine, locating himself/herself on the Earth's surface by taking instrument shots of the sun or stars and making necessary calculations, surveying and calculating the volume of material in dumps, carrying survey lines underground by shaft plumbing (cord or wire with attached bob is suspended from the shaft surface) and instrument shots taken on the bob at a shaft station, controlling by underground surveys and calculations, the driving and connection of underground passages on and between various levels, computing the volume of coal in portions of the mine from survey notes, and drafting maps of the mine workings.

Tailings. a. The gangue and other refuse material resulting from the washing, concentration, or treatment of ground ore. b. Those portions of washed ore or coal that are regarded as too poor to be treated further. c. Applied to sectional residue, e.g., table tailings, which is the residue from shaking screens and tables. d. The reject from froth flotation cells.

Tailings machine. A machine for sifting the tailings and collecting the gold from the detritus after it has passed through the washer.

Thickener. The concentration of the solids in a suspension with a view to recovering one fraction with a higher concentration of solids than in the original suspension.

Tipple. Originally the place where the mine cars were tipped and emptied of their coal, and still used in that sense, but more generally applied to the surface structures of a mine, including the preparation plant and loading tracks.

Top operator. A worker who is employed at surface jobs around the mine plant.

Tower crane. A swing-jib (crane with one horizontal boom on which there is a counterweight) or other type of crane mounted on top of a tower, the base of which may sometimes move on rails. These cranes are especially effective in congested sites.

Tram. a. A trip of coal cars or a single tramcar. b. Generally, to move a self-propelled piece of equipment other than a locomotive. c. A boxlike wagon of steel, running on a tramway or railway in a mine, for conveying coal or ore.

Trimmer. An apparatus for trimming a pile of coal into a regular form (such as a cone or prism).

Undercutter. In salt mining, an electrically driven machine somewhat like a gigantic chain saw. It has a long, thin horizontal bar, about which revolves an endless chain with cutting bits. The most common type is an adaptation of the shortwall coal cutter, a drag-type machine with continuous pick-filled chains to cut at the floor or bottom of the seam. It can make a rapid, continuous cut across the entire width of the face.

Underground mine. A mine that accesses a coal seam or other mineral through a shaft instead of removing the overburden to expose the seam [Vaught 2008].

Utility man. A worker expected to serve in any capacity when called on [Dictionary.com 2011].

Ventilation. Mine workings are usually subdivided to form a number of separate ventilating districts. Each district is given a specified supply of fresh air and is free from contamination by the air of other districts. Accordingly, the main intake air is split into the different districts of the mine. Later, the return air from the districts reunites to restore the single main return air current at or near the upcast shaft.

Wash plant. The place at which ore, coal, or crushed stone is freed from impurities or dust by washing.

Washery. A place at which ore, coal, or crushed stone is freed from impurities or dust by washing. Also called wet separation plant, washing plant, dense-medium washer, or efficiency of separation [Infomine Inc. 2010].

Weighman. One who weighs, measures, and checks materials, supplies, and equipment for the purpose of keeping relevant records [BLS 2010].

Wet plant operator. A person who works as a member of a crew performing any one or a combination of duties concerned with extracting cadmium, lead sulfate, and zinc oxide from dust recovered in Cottrell precipitators.

Yard. An area on the surface where mines store many of their supplies, such as bundles of roof bolts. These supplies are then sent underground or to the surface area of mining when needed [Vaught 2008].

www.ingramcontent.com/pod-product-compliance
Lightning Source LLC
Chambersburg PA
CBHW080238180526

45167CB00006B/2323